THE COMPLETE SAS SURVIVAL MANUAL

T0047156

Barry Davies, BEM

Skyhorse Publishing

Skyhorse Publishing books may be purchased in bulk at special discounts for sales promotion, corporate gifts, fund-raising, or educational purposes. Special editions can also be created to specifications. For details, contact the Special Sales Department, Skyhorse Publishing, 307 West 36th Street, 11th Floor, New York, NY 10018 or info@skyhorsepublishing.com.

Skyhorse® and Skyhorse Publishing® are registered trademarks of Skyhorse Publishing, Inc.®, a Delaware corporation.

www.skyhorsepublishing.com

10 9 8

Library of Congress Cataloging-in-Publication Data is available on file.

ISBN: 978-1-61608-282-6

Printed in China

CONTENTS

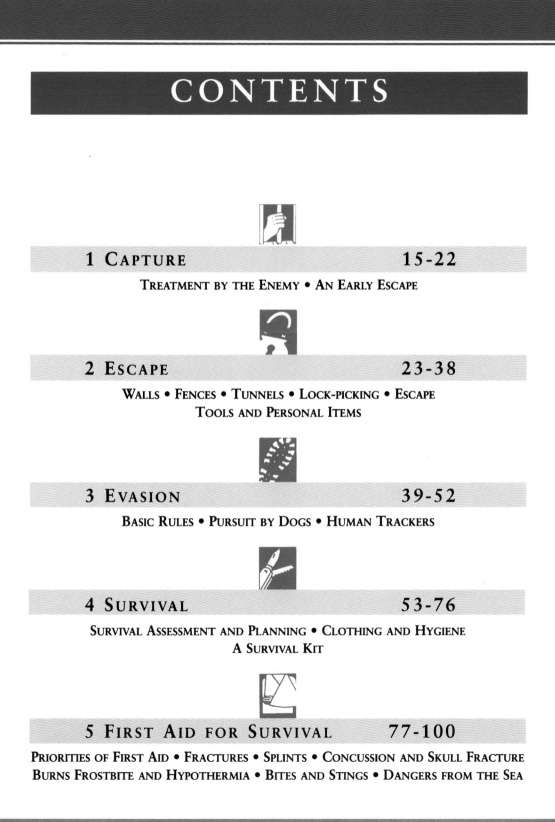

INTRODUCTION

This is my second book on survival; the first was published in 1987. In it I draw on a wealth of experience, all of which was gained through my involvement in survival training as a member of the SAS. During my 18 years with the regiment I was privileged to spend two years as a survival instructor at the International Long Range and Patrol School (ILRRPS) in southern Germany. There, while teaching survival techniques to pilots and special forces of many nations, I had the opportunity to expand the very principles of survival that I had myself been taught earlier. The best source of new ideas proved to be my favourite course: the one covering pilots' escape and evasion.

The moment I left the SAS I started a business manufacturing and marketing survival items for the general public. This new venture brought me into contact with BCB International Ltd, a Cardiff-based company with worldwide renown for its expertise in the field of survival. With BCB I have had the chance to devel-

op many innovative products that have helped the company to reach its present position as the world's largest survival equipment specialist. One such product was the new camouflage stick now used all over the world, including by the British Army. My connection with BCB remains strong, especially in the area of research and development.

This background, combining the SAS and BCB International, has led me to study survival techniques in every part of the world, from the steaming jungles of the Far East to the frozen wastes of Canada and the burning heat of the Sahara. And yes, I have tested most of the techniques described in this book. I can report that, depending on terrain, weather and the availability of equipment, many work well the first time, while others need practice. However, the main problem for the survivor is not one of equipment, and it is the same whether you are a soldier or a civilian. It is a matter of your will to do something about your predicament. Learn to deal with a survival situation by first of all recognizing it for what it is: nature's challenge.

The early chapters of this book cover escape and evasion, and although these are primarily of interest to military personnel, there is much here that the civilian survivor can benefit from. Dog evasion techniques may prove very useful, and if you find yourself contained by an electric fence you will be able to determine whether or not it is live. However, these two examples and the many other survival techniques described are offered to the reader in the hope that he or she will use them judiciously and appropriately. Anything that has a bearing on your safety and well-being deserves careful thought before it is undertaken. Moreover, in practising survival skills in the wild, be sure to treat the countryside with the respect due to it. For example, there is no need to pick wild plants unless you intend to eat them. Likewise, minimize cruelty to animals by setting traps and snares only when absolutely necessary to your survival. This means

not during training sessions, but only when the need is real. And always, before setting up a training survival camp, seek permission from the landowner if applicable.

The chances of survival are always good — as long as you stay calm and think clearly. To stay alive you need certain basic requirements — air, water, food, health and shelter. But to rise to the challenge of survival in a hostile environment you also need knowledge, imagination and practical skill.

Finally, never forget that should you find yourself in a survival situation, whether as civilian or soldier, help will always be available, and rescue agencies will come looking for you. Use the knowledge contained in this book to help them find you, and to remain alive until they do.

ACKNOWLEDGEMENTS

There are few books on the subject of Escape, Evasion and Survival and, to the best of my knowledge, none that contain all the up-to-date techniques and equipment described in the present volume. Although this is not an official publication of the regiment, it would be sadly remiss of me if I did not take this opportunity to acknowledge in full the experience I gained during my eighteen years' service with the SAS.

Additionally, the ongoing development of survival techniques and hardware by BCB International Ltd of Cardiff has helped me to provide the reader with the very latest information. Therefore I would like to thank the Managing Director, Andrew Howell, and all his staff, for their unceasing help and support. My sincere thanks also go to Lifeguard Equipment Ltd of Clwyd for their kind invitation to join the RAF Search and Rescue trials in 1994.

Putting this book together would have been impossible without the help and persistence of all the team at Siena Artworks of London.

Finally I would like to thank the brave people, both soldiers and civilians, who in the past have survived their own individual ordeal. It is only by expanding on their immense knowledge and experience that books like this can be written.

CAPTURE

'Survival situation' is a term that is often used very loosely. Basically, however, it implies that something, usually unplanned, has happened, so that you find yourself in a totally unknown and unexpected environment from which there is no immediate prospect of extrication. If you are a captured soldier, your life may be under threat. Your physical state and your exact location at the time of being taken prisoner will determine to a large extent your reaction to finding yourself in this kind of survival situation. Being captured by the enemy ranks as one of the most frightening experiences a soldier ever faces. The immediate fear of the unknown and the looming threat of death, or at best a severe beating, play havoc with the military prisoner's emotions. And the complex, seemingly insoluble nature of your predicament fills your mind with a sense of isolation and abandonment.

TREATMENT BY THE ENEMY

In modern warfare, even when the conflict takes place on a large scale, the number of captured soldiers is quite small compared with corresponding figures for the Second World War. As a result the treatment of prisoners of war has changed. In some cases these smaller numbers of men are nowadays treated to a more rapid and harsh treatment, while the typically greater ratio of guards to prisoners has tended to make escape ever more difficult.

However, the one fundamental thing on which any prisoner of war from the Western world can rely is that he will not be forgotten. The treatment of POWs is regarded as among the highest of priorities by any civilized government. And all steps are taken to establish contact, ensure the prisoner's well-being and finally administer his release and safe return home.

To be captured is not a dishonour – it is simply one of those unavoidable acts of war. Where men are sent into battle on land, or are engaged in aircraft sorties, operating deep behind enemy lines, the risk is always present. Although capture is sometimes inevitable, if there is even a faint chance of avoiding this fate, then that opportunity must be seized.

Being captured by the enemy is one of the most frightening experiences a soldier faces. The best insurance against this is to grab even the slimmest chance of avoiding capture.

The prisoner of war cannot expect to be treated in any set way by his captors, despite all the words and laws governing the treatment of POWs. Various factors will determine the prisoner's fate. For example, if a large number of prisoners have been taken, then the organizational demands of the situation may result in a lessening of ill treatment of the individual. But if an individual, say a pilot, is captured, then his captors may seek revenge, or at least use him as a symbol of this, by subjecting him to a beating. The prisoner's treatment will also depend on the professionalism of his captors. Professional troops normally act in a restrained and responsible way, whereas local militia are likely to be more crude in their approach.

The prisoner of war must be prepared to encounter some hostility from his captors, and the intensity of their reaction to him will be directly related to one or more of the following factors:

● Which side is winning the conflict
● The number of alleged atrocities carried out by forces fighting on his side
● A bombing mission by his forces that has or may have inadvertently killed innocent civilians.

This last act will be seen as a war crime and attributed to the individual prisoner. From the perspective of his captors, he is the enemy personified, and therefore responsible for all the actions that have been carried out by his own comrades.

AN EARLY ESCAPE

It is every soldier's duty to attempt to escape at the first opportunity. The closer the prisoner of war is to his own lines, the greater his chances of success. At such times he will know where his own troops are located, he will still be fit (unless he has been wounded), and he may well still have some of his equipment. However, a POW who has been captured close to the forward edge of the battle area must remember that the danger of being shot while trying to escape will be very high. Heavy concentrations of armed men in an acute state of tension will make escaping dangerous, and unless there is a good chance of success, it is unwise to risk provoking front-line combat troops.

Always watch out for the opportunity to escape during transit deep behind enemy lines. Escape is possible by all modes of travel: on foot, in a road vehicle, boat or train or by air. Stay alert, take advantage of diversions such as air strikes by friendly forces, or when the guards are sleeping. Even if the possibility of escape is not immediately obvious, the POW should collect any useful items and information that may aid his escape at a latter date.

You will find in this book a wealth of detailed information on practical escape, evasion and survival techniques, and with the help of these you can come through most survival situations. In recent years, great advances have been achieved in the design and manufacture of military

Evading capture when behind enemy lines demands an alert state of mind and a readiness to take advantage of diversions such as air strikes by friendly forces.

clothing, equipment and navigational aids. In addition, search and rescue techniques, and the accuracy and reliability of individual satellite communications, have improved dramatically.

The sophisticated technology of survival has an undeniably important role to play, but, above all, your chances of emerging alive will depend on your own capabilities as an individual. If you are captured, or if you are forced to survive on your own for any length of time, without the promise of imminent rescue, a number of potent psychological factors will come into play. It is important to understand just how powerful these psychological pressures can be.

PSYCHOLOGICAL EFFECTS OF CAPTURE

It is an unfortunate fact that our psychological reactions to disaster or a threatening situation often render us unable to make the best use of the available resources. Your first step therefore must be to control and direct your own reactions to finding yourself in a survival situation.

What are the psychological pressures confronting the soldier who is in a survival situation or is captured? We can describe them as the 'enemies of survival'; they include pain, fatigue, boredom, loneliness, and the effects of cold, thirst and hunger. Either separately or in combination, they work to induce fear in the individual. While everyone has some experience of all of them, very few people have known them to such a degree that they felt serious doubt about the possibility of actually surviving. And yet in a survival situation these feelings are essentially no different from those that we experience in less unusual conditions. They vary only in their severity and the danger that they present.

The first step towards taking control of your situation is to acknowledge the psychological pressures of capture and to understand how they affect your prospects of survival.

Unless he is trained to resist their insididious effects, the psychological pressures of capture can easily demoralize the prisoner of war.

PAIN

A normal occurrence in everyday life, pain is nature's way of making you pay attention to something that is wrong with your body. But nature also makes it possible to hold off pain even when it is severe – in order to diminish its debilitating effects – if you are so busy doing something important that you are unable to attend to the problem immediately. Pain may be made more bearable if you occupy your mind with plans and activities aimed at ensuring your survival. If, on the other hand, you do not attempt to combat it with your mind, pain will almost certainly diminish your ability to survive. Even when it is not serious or prolonged, it can get the better of you if you let it.

COLD

Suffering from cold is a much greater enemy of survival than it may at first appear. The obvious threat from cold is the physical damage it can inflict. Yet it is far more insidious than that. It numbs the mind as well as the body. In addition it both enfeebles the will and reduces the ability to think clearly. It can do this by such gentle, imperceptible stages that it is essential, before it takes effect, to adopt a positive attitude to resisting and guarding against it.

Cold has an adverse effect not just on the body but also on the mind. Before long, exposure to low temperatures makes it difficult to think clearly – a serious drawback when your life is in danger.

THIRST

Probably the best-known enemy of survival, especially in its extreme form, is thirst. Well before this stage is reached, however, thirst can begin to preoccupy the mind. But, like pain and cold, thirst can be kept in proportion. Indeed if it is not kept in check it can lay the sufferer open to the effects of pain, cold and fear. Your will to survive must be strong enough and it must be backed by confidence. That confidence must rest on the certainty that, if there is water to be had, you have sufficient knowledge to set about getting it. Of great importance, too, is the rational use of any stock of water you may already have. It must be used sparingly, but not so sparingly that you risk serious dehydration.

HEAT

The need to survive heat usually implies that the climate is dry and arid, as in a desert. Among the main problems in a hot environment is water – both preventing its loss from the body and acquiring a supply for drinking. Dehydration and heat exhaustion often occur together, weakening the individual to the extent of causing a total collapse. Dizziness, severe headache and an inability to walk are signs that emergency measures are needed. The first priority is to keep the body covered to avoid loss of water through sweating. The second is to ration whatever water is available.

Your body will adjust to a hot climate over 24-36 hours. However, bear in mind that in most desert areas the night-time temperature often drops below freezing point. If possible, rest by day and work by night.

HUNGER

The physical effects of hunger are obvious, especially if it is prolonged. Its additional hazard lies in its adverse mental effects, one of which is to lessen the capacity for ordered thought. Like thirst, hunger can lead to an increase in the survivor's susceptibility to the debilitating effects of cold, pain and fear.

FATIGUE

Even a moderate degree of fatigue has its own method of weakening the victim. It achieves this by creating a vague mental attitude in which it becomes increasingly easy not to care about what will or will not happen. Fatigue is another of the great threats to your survival, and all the greater because it can result from an unsuspected source. For it can be due to lack of hope, or of any real goal. It can build up from frustration, dissatisfaction or boredom. It may be unconsciously used as an avenue of escape from a reality which feels too difficult to contemplate. On the other hand, unknown reserves of strength can often be summoned, along with the will to go on – provided the dangers and sources of fatigue are recognized and fought against.

Prolonged exposure to heat causes several problems, the most serious of which are dehydration and exhaustion. Finding shelter and water without expending a great deal of energy is therefore a priority.

Your imagination and memories are your only friends when you are isolated. Planning your escape is a valuable antidote to the sense of isolation and may bear fruit.

The dread of torture and the possibility of imminent death can be countered by training the mind to accept fear and deal with it.

BOREDOM AND ISOLATION

These are two tough enemies for any prisoner of war, and they tend to occur together. Their toughness derives from the fact that they are usually unexpected. Dealing with waiting when there is nothing happening can play a very significant part in survival. Many hopes and expectations may be raised, only to be dashed to nothing. At night, for example, you may have to stay still, quiet and alone. This is the time when you become the target of a combination of boredom and loneliness. At such moments they will sidle into your consciousness unless you defend yourself against them.

The best antidote is to talk – to yourself if necessary – and make plans for the future. Talk about the future that awaits you after your escape. Devise problems to keep your mind exercised and occupied. Active, positive thinking leaves no room for boredom or loneliness.

FEAR

Fear is an entirely normal, and indeed necessary, emotion. It is the instinctive reaction of anyone facing a threat to life or limb, or imminent torture. In survival situations behaviour and reactions are always influenced by fear, and, through them, so are the prospects for survival. There is no advantage – indeed there may be a strong disadvantage – in attempting to avoid fear by denying the existence of danger. Acceptance of fear as a natural reaction to any threatening situation will produce two immediate and positive benefits.

First, you will be able to dismiss the fear of being afraid, which is often a burden in itself. True courage is often to be found in people who freely admit to fear, and then go on to do their best in the circumstances they face. Second, you will find yourself more likely to be able to carry out considered rather than uncoordinated actions. With your mind both clearer about the danger you confront and more able to command action, you will recognize that there is always something that can be done to improve the situation. The key is: never lose hope.

2

ESCAPE

This chapter is not intended to make the reader an expert on escape. Its purpose is rather to help anyone who is facing the risk of becoming a prisoner of war to apply his mind to the possibilities open to him in that situation. Each subject covered is intended to act as a catalyst that will spur the user into positive action should he find himself in captivity.

All prisoners are normally confined by one or more types of structure, and it is these that need to be analysed before a serious escape attempt can be planned. The first problem to solve is: Do you go over, through or under? The answer lies in the nature of the structure that confines you.

WALLS

The five materials that are most commonly used in the construction of walls used to contain prisoners are: brick, stone, block, timber and reinforced concrete. Each of these presents a different degree of difficulty for the intending escaper.

Scrape away the mortar and remove a single brick to break the bond. If this will take more than one session, mix the removed mortar with water or urine and refill the hole after working. This will hide your efforts from the guards.

BRICK

A brick-built wall is the type that the POW is most likely to come up against, and fortunately these are among the easiest to break through. All brick walls get their strength from their mortar bond. Break this bond and you break the wall.

Begin by selecting the position where you wish to exit. Starting at the middle, remove all the mortar from around a single brick. This is best carried out by continually scraping away at the mortar with an improvised chisel. This task should be undertaken only when you are

completely isolated, with no guards present. It may take several days of sessions to completely remove the first brick, which is always the most difficult. At the end of each session, collect all the powdered mortar scrapped from between the bricks, and wet it with a little water or urine and replace the mortar carefully to disguise your handiwork. (Adding soap to the water will aid this binding process.) This should prevent any of the guards noticing your work, although it is safer to select, if it is possible, a section of wall that is obscured by some other object, for example a bed.

Once you have removed the first brick, the bond is effectively broken, although it may be necessary to remove several bricks in this way before the rest are loose enough to be broken out by hand.

BLOCKS

Many buildings constructed in recent decades are built of large concrete blocks. These should be treated in exactly the same way as bricks, although they are more difficult to remove in one piece. However, the possibility of breaking the block, especially if it is of the type that are hollow inside, is very high. Walls constructed of single hollow blocks can be smashed through in a very short time using a home-made hammer and chisel. A short piece of steel plumbing pipe may, as well as providing a chisel, also serve as an improvised sledgehammer. Breaking through blocks is a good escape method if noise is not a limiting factor.

Hollow blocks can be broken through easily with an improvised hammer. This is an ideal escape method when noise is not a problem.

A stone wall will probably be difficult if not impossible to penetrate. Look instead for a weak point such as a door or a window frame.

STONE

Normally found in older buildings, stone walls are generally extremely difficult to penetrate because of their thickness. Although the same basic principle is applied as in escaping through brick walls, it usually takes much longer to break though the several layers of different-sized stones. In such a situation

it is a good idea to examine the other openings in the room, in particular windows and doors If the building is indeed old, these may well have deteriorated to such an extent that they can be broken away from their fixings without great difficulty by using the improvised hammer and chisel.

TIMBER

Buildings made of timber, except those in which solid logs have been used, do not pose a great problem. A length of metal water-pipe flattened at one end can be inserted between overlapping joints to force timbers apart. Panels from which the nails have been removed can offer good escape holes. Escape via the roof of a

A wooden prison hut offers various possibilities for escape. If you cannot break through the walls or the roof, you may be able, if the hut stands over bare earth, to escape through the floor or to dig a tunnel.

wooden building should also be considered. So should the possibility of tunnelling, as some timber buildings erected for temporary use stand on earth foundations.

REINFORCED CONCRETE

Buildings made of reinforced concrete pose a major problem. However, this material is normally found only in buildings with special security arrangements and in foundations, including cellars. In such cases escape through the walls is virtually impossible. Look for other means of exit, such as doors, windows, air vents or sewage ducts.

FENCES

Although fences in civilian use are of various types of construction, those that the prisoner of war will confront are almost always of wire and are either used as a temporary enclosure measure or as a secondary perimeter barrier. All wire fences need to be inspected carefully, to determine their overall construction and the thickness and type of wire used.

The construction of the fence is a crucial factor, for some manufacturing methods allow a certain number of links to be cut so that a large section of the fence can be collapsed. The thickness of the wire is also important, especially if the prisoner intends to cut the wire. Additionally he may wish to climb the fence, and he must be sure that it will bear his weight. The type of wire will influence his decision on how to tackle the fence. For example, if razor wire is present, then the POW will need to wear protective padding if he intends to scale the fence.

As with walls, study the nature of the problem carefully before deciding on how you will try to escape. The construction of the fence should help you to make your mind up as to whether it will be best to go over, under or through it.

LINK AND MESH FENCES

Most wire fences are constructed by weaving metal links together. Cutting the links in a set pattern will reduce the number of cuts and shorten the escape time. However, rigid mesh fences tend to be used in more modern prisons and these are best climbed using a home-made claw grip, which hooks over the fence's top.

Metal fences that can be climbed are often topped by a secondary barrier which may incorporate razor wire, barbed wire or rolling drums. The first two can normally be crossed

DROPPING

If your escape requires a drop from a height, you can reduce the risk of injury by using the following technique. Lower yourself to full stretch, then let go and relax as you drop, keeping your feet and knees together and your elboes tucked into your chest. On landing, roll forward in a ball. Any drop of more than twice your own height will be dangerous.

If a fence is topped with razor wire or barbed wire, use a heavy material to protect yourself as you climb over.

Cut the right links and you will be able to penetrate the fence.

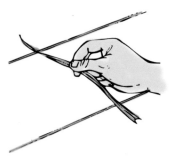

Advance the blade of grass slowly towards the suspected electrified wire until you can feel a tingle from the current.

by employing an improvised protective cloak. A thick but pliable material, for example carpet or heavy canvas, can be fashioned into a Batman-style cloak which is far less likely to get snagged than your clothing as you climb the fence. To protect your body from the hazardous wire as you climb over it, it is a simple matter of throwing the cloak over your head and releasing the fastening at your neck. Rolling drums need to be secured by tying before they can be crossed.

ELECTRIC FENCES

It is seldom that electric fences are used to house POWs, but if you are not sure whether a fence is electrified it is a wise precaution to check. This is done by simply placing a small blade of grass against the fence while making sure that you do not actually touch the fence with any part of your body. Hold the blade of grass in your hand and touch the tip to the fence. If you can feel nothing, advance the blade of grass by slowly bringing your hand closer to the fence. If, by the time your hand is within half an inch of the fence, you can still feel no tingling sensation, the fence is not live. It is important to note that in some modern installations

the fences are electrified by intermittent pulses. There-fore if you have any suspicion at all that one of these is in use you should prolong the time you hold the grass to the wire. The pulse cycle may be short, with brief intervals between pulses.

TUNNELS

A very popular way of escaping during the Second World War, tunnelling relied above all on the availability of two important assets. The first was sufficient manpower to dig the tunnel and then distribute the excavated earth. The second was sufficient time, for tunnels of the length that was usually required weeks or even months to dig and shore up.

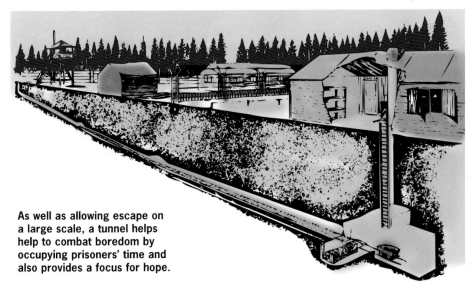

As well as allowing escape on a large scale, a tunnel helps help to combat boredom by occupying prisoners' time and also provides a focus for hope.

In modern warfare this method of escape from captivity is still possible, but since both the factors listed above are more likely to be absent nowadays, it is best to restrict tunnels to short ones under fences that can be dug by one person in a single night.

In modern warfare this method of escape from captivity is still possible, but since both the factors listed above are more likely to be absent nowadays, it is best be dug by one person in a single night.

AN INVENTIVE ESCAPE

The major escapes and attempted escapes of the two world wars have been extensively documented, but since the Second World War some ingenious, and in many cases dramatic, examples of both have taken place in various parts of the world. These were carried out by people who were desperate, and in some instances they risked not just their own lives but also those of their families.

During my research for this book I read about some forty different escape attempts, ranging from the very straight-forward to the highly dangerous. But perhaps the most remarkable of these was the man who, with his wife and children, escaped the communist system of the former East Germany by flying to the West in a hot-air balloon. The family made one abortive attempt, but the second time, as they held on tight to the improvised platform suspended from the simple home-made balloon, they were lifted silently and, by very good fortune, undetected, over a strip of land-mined ground and formidable border fences.

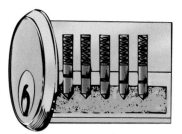

In a typical pin-tumbler lock the inner barrel, when turned by the key, releases the lock. The barrel is held in position by a series of split pins seated on springs.

LOCK-PICKING

During the Second World War prisoners of war opened many doors with skeleton keys and by picking the locks. It is said that the inmates of Colditz enjoyed a free run of the supposedly secure German POW castle thanks to these skills.

The tools used in picking locks can easily be improvised and the basic principles of the art are fairly straightforward. The problem lies in mastering the practical skills. It can take many years to perfect the fundamental techniques of lock-picking, and constant practice is needed to acquire the necessary 'feel' for a variety of different locks.

To any reader who might consider it irresponsible to publish information about lock-picking, I offer the reassurance that it took me about two years to learn even the basics of the art, and that was with expert instruction and constant practice. Moreover, if a person wishes to learn lock-picking seriously, there are in the UK and other countries professional bodies that provide legitimate instruction in the craft.

LOCK DESIGN

Most door locks manufactured over the past twent
years are of the pin-tumbler type. In its basic form
this is a very simple locking device. A series of small
pins fit into the inner barrel of a cylinder. The pins,
which are the same length but split at differen
points, are forced into recesses within the inner bar
by a small spring. If an appropriate key is inserted, the
pins are brought into line so that the split in each meets
the outer casing of the inner barrel. This allows the
inner barrel to turn freely within the casing and thus
release the lock.

Any method of aligning the pins in this way and so
turning the inner barrel will open the lock. One of two
methods is used for this purpose: raking or picking.

When it is inserted, the key brings the split portion in each pin level with the inner barrel and the outer casing. This allows the barrel to turn and the lock to open.

RAKING

Although the techniques of raking and picking differ,
similar tools are employed in both. These implements
– here called rakes but also known as ball picks, large
diamonds, hooks and tension bars – are not normally
available in most countries, but a set of lock-picking

Always hold a lock pick or rake with no more pressure than you would a teaspoon.

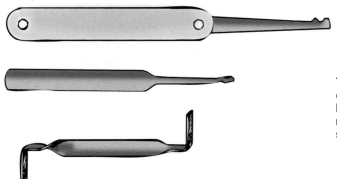

The lock-picking tools shown on the left are, from top to bottom: a double ball pick, a rake and a double-ended tension bar.

tools is fairly easy to improvise. The rake is simply a
flat strip of hardened metal that has its end shaped to
fit into the lock and advance the pins on their small
springs to the required position. In addition a tension
bar is required. This is also a flat strip of metal, which

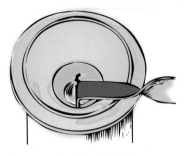

Use a tension bar which is wide enough to turn the barrel but which leaves enough room for you to insert a rake or pick with ease.

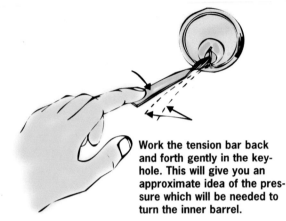

Work the tension bar back and forth gently in the key-hole. This will give you an approximate idea of the pressure which will be needed to turn the inner barrel.

is inserted into the mouth of the barrel in order to apply pressure, thus helping to both turn the barrel and seat the pins.

Many different designs of lock-picking tools are used for different functions in the process, but the two described above are sufficient for all needs.

The technique of raking imitates the key's action. The tension bar acts as the key shaft, while the abrupt raking action aligns the split pins so that the inner barrel turns and the lock opens. Raking is the quickest method of opening a lock, and it is easily done provided the pin sizes do not change suddenly, as in the combination lock. The lock should be clean and free from grit and dirt before any attempt is made to open

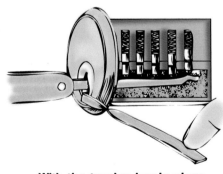

With the tension bar in place, insert the rake to the far end of the inner barrel.

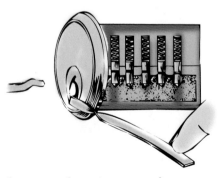

At the same time as you apply slight tension to the bar, snap out the rake smartly.

LOCK QUALITY

The ease with which a lock can be opened depends on three factors. Undeniably important among these is the effectiveness of the tools you use, and here, as with most practical tasks, inferior quality should be avoided. The other two factors concern the lock's design. The first is the pins' length and position and the second is the quality of the lock. Cheap locks are, in almost every case, easier to open than expensive ones. They are generally poorly constructed in that a much greater clearance is allowed between the barrel and the body, in order to make assembly easier and cheaper. Other common features of inexpensive locks are poor barrel alignment and oversized pin holes, both of which faults make them very easy to pick.

it. The best way to achieve this is to blow hard into the lock before you start work.

Basically, raking is a matter of inserting the pick to the rear of the pins and swiftly snapping it outwards, running the tip over the pins in the process. Before this can be done, however, a tension bar must be inserted into the bottom of the key-way and used to apply slight tension to the lock's barrel. This tension, which is applied in the 'unlock' direction, should be just sufficient to turn the barrel once the pins are seated, but not so strong as to bind the pins against the barrel. Recognizing when the tension is right is the basis of all good lock-picking. If it is too great the top pins will bind and the sear line will not allow the breaking point to meet.

If raking fails, on the final rake hold the tension bar in place. Listen to the pins falling back into place. A soft sound means a pin at the back of the lock is sticking.

LOCK-PICKING AND THE LAW

Readers will be aware of the existence of laws concerning the crime of 'breaking and entering' premises. However, it should be noted that unless the holder practises lock-picking as a legitimate professional trade, it is illegal to carry in public tools intended for use in picking locks. The guide to lock-picking techniques which appears here is to be regarded exclusively in the context of escape by prisoners of war.

If it is too light the pins will simply fall back into the locked position.

When raking it is usually necessary to repeat the above operation several times. If the barrel does not turn on the fourth try, maintain the tension with the bar, place your ear to the lock and then slowly release the tension. If you hear the 'pitting' sound as the pins fall back to rest, you have applied too much tension, whereas if you hear nothing then you need to apply more tension with the bar.

PICKING

Lock-picking is similar to raking, but requires much more skill because the pins need to be seated individually. Looking at the back of the lock, feel for the rearmost pin and gently push it up.

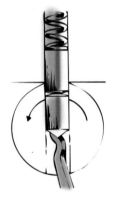

To deal with stubborn pins, rake the lock while applying light tension. Insert the pick fully (above left) and withdraw it until you feel a problem pin, which will be stiff. Reduce tension on the barrel minimally and waggle the pick back and forth (above right). As each problem pin is seated, the inner barrel will move a tiny amount.

This should move the barrel fractionally. Working towards the end of the lock, seat each pin in turn until the barrel is released. One swift rake, followed by picking, is sometimes the easy answer.

When the lock is a padlock the task of opening it can sometimes be made easier by sharpening one end of the pick to a needle point. When this point is forced all the way to the rear of the lock until it hits the rear plate, the pick will grip the metal. Also, try forcing the plate down or up, as this will sometimes release the lock without the need for raking or picking.

In some padlocks the pins can be bypassed by using a sharpened pick to open the gate in the key-way.

LOCK PICKS

Various easily acquired items, including safety-pins and tempered wire flattened at one end, can be pressed into service as lock picks. Alternatively, picks can be cut from hard plastic, while for those with access to a machine shop, a good lock-picking set can be made from a set of heavy-duty feeler gauges.

ESCAPE TOOLS AND PERSONAL ITEMS

One of the main difficulties of being confined as a POW is boredom. You can alleviate this problem by occupying your mind with the task of devising and making escape tools or items that will improve the conditions in which you are living. For example, even a small stick with the end crushed and splayed out will serve as a toothbrush, and feeling that you are looking after your body is very important in maintaining a sense of well-being. As well as making new items, be sure to look after the ones you already have, including your clothes and boots; the condition of these may well prove an important factor in escape and evasion.

Listed below are a few tried-and-tested ideas for escape tools and personal items that will improve a period of captivity. But no matter what items you chose to make, always give free rein to your imagination when devising them. Just as much as practical skill, ingenuity and resourcefulness are the keywords here.

Food Any food that you may receive in captivity should be eaten. However, if is at all possible and an escape is planned, some food should be reserved. Sugar,

Dissolve sugar with a little water over a fire and allow to cool. This will provide a supply of energy-giving bars for your escape rations.

Survival tip
On all SAS survival courses the first item of equipment each participant acquires is a billycan.

for example, can be kept and turned into a solid bar of energy by simply adding a little water to dissolve it and then heating it, to produce a firm block. All foodstuffs with a high sugar or salt content will keep for a considerable length of time and so are ideal for storage as escape rations.

Map Get a map, make a map or steal a map. If you intend to escape, this is one of the first things you will need. If a map is not available, make one by drawing it on the inside of your coat, jacket or shirt. If there is no alternative, retain a map in your mind.

If you have succeeded in escaping, look out for any type of map, regardless of the scale or size. They can be found in vehicles, telephone boxes, on dead soldiers and in many other places.

Chisel In this situation a chisel can be almost anything metal. Its main function will be to scrape the mortar from between bricks in order to escape

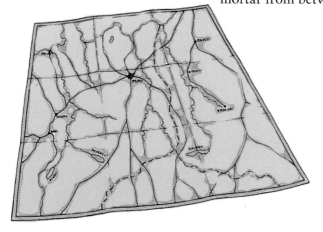

You need to make some escape tools. A chisel, for example, can be made from a length of waterpipe. Leave a nut on one end and flatten the other end.

A map is essential, particularly if you are deep behind enemy lines. In the chaos of war one can usually be found on a dead soldier or in a vehicle or building.

through a wall. Sources range widely, from a metal boot tip to a kitchen utensil.

Compass Escaping is just the first step towards regaining freedom. Next comes evasion and travelling to safety, and here you will need a compass. The one illustrated overleaf was constructed in the POW castle of Colditz, in Germany, during the Second World War. All that is needed is a magnetized needle and some way to balance it so that it can point north. This can be achieved by placing the magnetized item on water, or suspending it in the air on a fine thread.

Bones All animal bones are useful for making escape and survival tools. The sort most likely to be obtained while in captivity are the small ones found in meat dishes. With minimal adaptation these will provide needles for sewing, buttons for clothing and handles for other home-made tools.

USING THE WEATHER

Some escapes are planned, while others take advantage of a sudden opportunity or a change in the circumstances. The weather is often the determining factor in the timing of opportunistic escapes. A dark, stormy night, for example, may not seem ideal but the noise of heavy rain and thunder and lightning is likely to divert the guards' attention and a sustained downpour and poor visibility can aid concealment. Also, if the storm is fierce it may well knock out the local power supply.

Above is a neatly concealed compass from Colditz. A magnetized needle, right, acts a compass if suspended or set on a floating leaf.

Tyres Vehicle tyres of any type are an excellent source of material for making a number of items, including shoes, belts and, in an emergency, fuel for a fire. A burnt tyre will supply a large amount of wire for traps and snares. But note that burning a tyre produces a huge plume of black smoke which will be visible over long distances, particularly from the air.

Tin can An empty tin can will serve the survivor well as a billycan. This is a multi-purpose vessel which can be used as a mug and a mess-tin, for storing and carrying food and water, and as a kettle. A larger tin can makes a cooking pot.

Bark The bark of certain trees can be used to make clog-like shoes and eating and drinking vessels and may be suitable for making other escape items.

Improvised tools improve your situation. Needles and buttons will repair clothing. A bone holds a piece of metal to make a knife. Branches will provide forks for eating. A stick is a useful cooking aid and wood will make cutlery, including chopsticks.

3

EVASION

The division here of evasion into short-term and long-term is not simply a matter of the duration but also of the circumstances of the evasion attempt. In short-term evasion the soldier will have become separated from his unit or have been captured in the battle area and escaped. In either case his immediate concern is evading the enemy. Since he is close to friendly forces, evasion may last only hours or, at worst, days. In this short period of evasion, unless wounded, the soldier is likely to remain fit, and he will have a good sense of his position since he will not have been able to travel very far. However, stealth and constant alertness are essential here.

By contrast, the long-term evader, for example a downed pilot or an escaped prisoner of war, will face many problems, and his attempts to return to safety may last weeks or even months. During this time he faces several pressing demands – gathering food, locating water, staying warm and healthy, and travelling considerable distances through hostile terrain.

BASIC RULES OF EVASION

The success of an evasion attempt depends on adherence to four basic principles:

1 **Preparation**
   This means devising a workable plan before the escape takes place and preparing yourself both mentally and physically. Decide on the basic direction and then the route you are to follow. Consider as many of the 'what ifs' and possible problems as you can think of.

2 **Escape and survival equipment**
   Conserve durable foodstuffs for use as evasion

rations. Never throw away any other items that might prove of practical use in escape or evasion.

3 Observe the basic rules of military covert movement, which are as follows:

Make full use of the vital elements of camouflage and concealment. If possible select the route which offers the best cover. If you must travel in the open, move only at night or when you are completely sure it is safe.

4 Never take chances.

Stay alert. Be patient and be confident.

PURSUIT BY DOGS

For any escaping prisoner, the threat of dogs comes from two directions. The lesser risk is from some breeds of domestic dogs, which, in performing their role as watchdogs, can compromise a POW's position. However, a far more serious danger is detection, pursuit and capture by tracker dogs under the control of trained handlers.

Man has used dogs for military purposes, as well as for tracking criminals, for thousands of years. The Egyptians, Romans and Huns, among others, all relied on the services of both guard and tracker dogs. There is no reason to assume that the tactics that escapees used then to evade dogs have changed greatly.

CHOOSING A PURSUIT DOG

Dogs used for pursuing and apprehending escaped prisoners must fulfil certain physical and temperamental requirements.

PHYSICAL CHARACTERISTICS
Height at shoulder: 56-66cm (22-26cm) Weight: 20-45kg
(45-100lb)Speed: 40- 48 kph (25-30mph) over short distances

TEMPERAMENT
Intelligent, courageous, faithful, adaptable, energetic.

German Shepherd

Labrador

Certain breeds of dog are particularly suitable for use in tracking humans. They include the Dobermann, German Shepherd (Alsatian), Labrador, Mastiff and Rottweiler.

THE DOG'S SENSES

In its day-to-day activities the dog relies very little on sight. Nor is there any evidence that its night vision is any better than a human's, although its low position may help, as more of what it sees is better defined. However, by day or by night a dog's attention is drawn by visible movement and if its interest is aroused in this way it will investigate further with its ears and its nose.

Since a dog's hearing is twice as acute as a human's, it can be attracted by a noise not heard by the handler. The distance at which sound is received is largely determined by the weather, particularly wind and rain.

The dog's main sensory advantage over a human lies in its sense of smell, which is widely estimated to be between seven and nine hundred times more acute. A dog's nose can track microscopic traces of substances or their vapour lingering in the air, on the ground or when they have been in contact with other objects. A dog can also detect minute changes on the ground in the 'scent picture'.

THE GUARD DOG

Guard dogs are normally employed to detect intruders, locate them and detain them, thus protecting both the site and the handler. They can be employed in several different ways:

- Loose in a compound
- On a running wire
- On a lead with their handler.

THE TRACKER DOG

Tracker dogs are employed to find and follow the scent of a man's progress on foot. They concentrate on ground scent, unlike guard dogs, which work primarily on air scent. Normally a tracker dog will follow the freshest scent, but the achievement of the best results depends a great deal on teamwork between the dog and the handler. While the dog is trained to follow a distinct track, it is up to the handler to make sure that it is the correct one.

DOG EVASION

A dog which has spotted a moving man may lose interest if the man 'freezes' and shows no aggression. However, when a dog is in hot pursuit and closing on you there is not much you can do but defend yourself. In a delayed pursuit, even if the delay is very short, there are several countermeasures that can be taken.

THE SCENT PICTURE

The olfactory information in a particular site may be divided into two components: the air scent and the ground scent. Together these comprise the site's 'scent picture'. A trained dog analyses both elements thoroughly in its investigations.

Air scent consists mainly of body scent, clothing, deodorants, toiletries and the chemicals used in washing clothes. Of these scents, a dog is most able to detect the body scent given off by a human. The total amount of this in the air will depend on that person's constitution, activity and mental state. As a POW runs along, this scent is suspended in the air for a short while before dropping down to become ground scent.

From the dog's point of view, ground scent consists of two 'pictures'. The first of these is the body scent a person leaves and the other is the results of the disturbance he has caused as each foot hits the ground. His footfall unavoidably crushes vegetation and insects, and breaks the surface of the ground, each of which releases a vapour. Ground scent can last up to 48 hours, or even longer.

The main factors favourable to promoting ground scent are moist ground conditions, humidity, light rain, mist or fog, vegetation and forested areas. Unfavourable factors are arid areas, sand, stone, roads and urban streets, high winds and heavy rain.

The main recourse is to increase the distance between yourself and the dog. This can be done in various ways:

● Running steadily
● Climbing up or jumping down vertical features
● Swimming or wading across rivers.

Other useful tactics are:

● If you are in a group, split up to frustrate the dog's sense of single-minded purpose.
● Run downwind so that the dog loses your scent.
● Try to confuse the handler.

One ploy that will usually confuse the handler is to cross a river, walk some two hundred metres downstream and then cross back over. If this action is repeated several times, the handler will think the dog has lost you, and call the dog off, whereas what you have done is simply to confuse the handler.

REPELLING AN ATTACKING DOG

An attacking dog will attempt to paw down any barrier placed in front of it, so using a strong stick to bar its path may give you some temporary protection. Normally the dog will try to bite, to 'lock on' to you. If it is clearly intent on doing so, offer it an arm padded with an item of clothing. Once the dog has taken a grip, stab it in the chest or beat it on the head with a rock or stick. Make sure that whatever you do, the effect is permanent, otherwise the animal will just become even more annoyed.

If the handler is not present and you have no other weapon, try charging directly at the dog screaming and with your arms outstretched. Given the size of a human relative to that of a dog, and the sudden unexpected nature of the attack, the dog may turn tail. A dog's confidence and security can be weakened very quickly by challenging it in this way.

Never use chemical substances or pepper to put a dog off the scent, as this only improves the scent

> **Survival tip**
>
> If a dog is charging at you, try to break its momentum, as it relies on this to knock you to the ground. Stand, clearly visible to the dog, next to an obstacle such as a tree or wall until the dog is a metre or two away, and then, at the last second, move rapidly behind the obstacle. The dog will be forced to slow in order to turn, losing much of its momentum.

An arm cushioned by clothing will engage the dog so that you can kill it.

picture it receives. If you are cornered by both dog and handler, it is best to surrender unless you are armed. Killing the handler's dog will most likely bring severe retribution once you are recaptured. If you are fairly fit, it may be worth trying to outrun them, since the dog is only as fast as its handler.

HUMAN TRACKERS

It is not just dogs that are used to track down prisoners – in some cases expert human visual trackers are employed. Also known as 'scouts', trackers have been used since human conflict began. The skill grew out of our ancestors' desire for meat. The limited range and accuracy of early man's weapons forced him get close to animals and birds in order to kill them, and to do this he had to learn how to observe and approach unheard and unseen. As man turned his aggression on his fellow human being, the skills of the tracker of prey were employed to locate the enemy.

Visual trackers have been used with great success ever since those early times, a recent notable example being the Vietnam War, where their task was to locate, follow and gain information on the enemy as he moved through the dense jungle. Trackers are also used to track and recapture escaped prisoners of war. Although the tracker's main sense is visual, he may in some cases use a dog to assist him.

To avoid the dangers posed by visual trackers, the escaped prisoner must understand the principles they use. In this way he will be able to use countermeasures to avoid recapture.

SIGNS USED IN TRACKING

The tracker relies on a variety of tell-tale signs (in this context often referred to in the singular, as 'sign') left behind by a human passing through an area, to detect the presence and direction of his prey.

Temporary sign In this category are the unavoidable marks left behind on the ground, such as disturbance

> **Survival tip**
> My own extensive experience of Escape and Evasion exercises has convinced me that when attempting evasion it is always best to assume that you are being hunted.

To the tracker, a boot mark in wet ground is a definite, if temporary, sign of human passage through an area.

Top sign occurs above knee height – indicated here by a dotted line – and ground sign occurs below it.

of the earth and of leaves and sticks, dead insects and the disturbance of wild life. All such signs are temporary because after a short time, weather and the growth of vegetation will obliterate them.

Permanent sign Marks made by cutting or breaking (of branches, for example), dropped items and man-made objects left behind, are the main kinds of permanent sign relied on by the tracker.

Sign falls into two so-called 'visual ranges':

Top sign Generally found in vegetation which is growing above knee height, top sign is made when a human passes through undergrowth, causing a visible disturbance. The larger the group of people the tracker is following the larger the top sign. Consequently a tracker can follow a large top sign more rapidly than he could when tracking an individual.

Ground sign The disturbance of the track or path itself that the tracker is following yields ground sign. The main examples are young growing plants trodden down, footprints in soft soil, scuffing of fallen leaves, skid marks left by people climbing up or down hill. As with top sign, the larger the group being tracked the more evident the ground sign.

Examples of temporary small ground sign include:
- Broken, disturbed or squashed sticks and leaves
- Freshly disturbed earth and worm-casts
- Sand on leaves, rocks etc
- Shadow or shine on leaves
- The absence of cobwebs (easily visible as they are covered by morning dew).

Examples of permanent ground sign include:
- Man-made items such as confectionery wrappers, string, clothing
- Cut branches, sticks or vines
- Large skid marks
- Camp-fires
- Human waste products.

FACTORS AFFECTING TRACKING

Various factors influence the ease with which a tracker can locate and follow his prey's trail. The most obvious of these is the size of the party being tracked, since it is easier to follow the spoor (footprints) of a group than of an individual). The terrain is also important. It is easier to track through jungle than mountainous areas. Sun and rain also have an effect on tracking in that they slowly erode the spoor, and over a longer period vegetation grows up, erasing the disturbance caused by human presence.

From the spoor the tracker will identify the person or persons he is pursuing. Sooner or later he will find a print that will 'tag' an individual and if he is lucky enough to find several prints in close succession (in soft ground) he will be able to determine the speed of his quarry. The faster the quarry the longer the stride.

> ## THE SPOOR
>
> **The spoor – the footprint or footprints of the quarry – provides the tracker with several vital pieces of information. These include the number of people being hunted; the direction of the quarry; the speed of the quarry (this also indicates whether any injury has been sustained); the type of footwear worn; the intention of the quarry (it usually takes several days for a pattern revealing this to emerge).**

TERRAIN AND TRACKING

The type of terrain found in the tracking area brings certain advantages and disadvantages to both the tracker and the quarry. The following are the principal examples of the influence of different terrains on tracking and evasion.

Grassland If the grass is high, say above half a metre, the trails will be relatively easy to follow, as trampled-down grass will indicate direction. However, short, springy grass will return to its original position fairly quickly. Other problems for the escaped prisoner who is crossing grassland are that the grass presents a colour contrast when trodden down; he leaves a trail when the grass is wet with dew; and when the weather is wet he will deposit mud on the grass.

Rocky and desert country Tracking through dry types of terrain is difficult but not impossible. Rocks

A tracker looks for signs that can only have been left by a human. The remains of a camp-fire will tell him that you have the means to light one and have rested. If you have emptied your bowels nearby he can detect whether you are eating solids. The golden rule is: Never leave sign.

The tracker can easily distinguish between grass or other vegetation that has been trampled by human feet and grass which has been trodden down by animals.

In sandy ground where there is little rain, footprints can remain for many days.

can be disturbed. If the weather is favourable, distinct spoor can remain for days in sandy desert. Among the signs that the escaped prisoner must take care not to leave are rocks that have been rolled over (these are evident as they are darker on what was the underside). On sandstone, boot marks will show up as darker in colour, while on lava the marks will have a whitish appearance.

Any large sign left behind on a rocky or desert surface will be spotted without any difficulty by a tracker. Pure desert areas will preserve a spoor that remains visible over a distance of many kilometres. Hard sand will produce a clear footprint, whereas in soft sand the footprint will be sufficiently deep to leave a shadow.

Rainforest The jungle offers many ways for the tracker to identify his quarry, since the undergrowth is very thick and such areas are generally criss-crossed by small streams and rivers. Tracking is assisted by observing signs specific to jungle and forests; these signs are therefore to be avoided wherever possible by the evader. Disturbed leaves on the forest floor are a clear sign of human presence (animals don't leave the same sort of clues). In particular, dead dry leaves are brittle and so are easily crushed under foot. If the undergrowth is very thick, the quarry will be forced to make a hole through it – another sign that is very visible to a tracker. Other tell-tale signs are boot impressions left on the soft soil of stream banks, scuff marks made by climbing over fallen, rotting logs, and, in the early morning, broken cobwebs found above the head height of animals frequenting the vicinity.

Flowing waters and marshy areas Sign left behind in wet areas may be temporary, but it is usually sharp. Such indications of human presence include footprints on the soft banks of flowing water, mud on rocks at the water's edge, and discoloration of the water from mud being stirred up on the river-bed.

A footprint indicating a worn boot or shoe heel can identify a particular individual and 'tag' him for the tracker.

ESTABLISHING DIRECTION

Once the tracker has established the spoor he will soon deduce its direction. When he has done this it is not necessary for him to continually look for sign. The lie of the terrain will indicate the direction of his quarry. When we enter a house we normally, and logically, go through the door. The same applies outdoors: if we are surrounded by trees and thick scrub with a natural opening roughly on our line of march, we will pass through it. Occasionally the quarry will have climbed over a rotting log or crossed a clear soft patch of ground, leaving a 'confirmed sign' of his direction.

Marshy ground produces sign that is clear for a short while but soon disappears. But if the tracker finds it he can establish roughly when the print was made. Footprints will even remain for a short time in the mud of a river-bed and, depending on the rate of flow, the water can stay cloudy for as long as two hours.

CAMP-SITE REMAINS

More information can be gained if the tracker locates an overnight camp-site or a rest area. By studying this he can glean vital information. The nature of the indentations or shelters will help indicate the size of the group. Camp-fires and food scraps may determine the quarry's physical strength. The contents of rubbish holes and latrines will also reveal vital information on the size of the group being tracked.

The tracker will immediately know if the spoor has gone cold. At this stage, if he is confident that he is still following the right direction he may do a 'cast'. The first cast will be visual, and is used to establish if the quarry had a choice of opening on the natural line direction, and to check this first. If no sign is found the tracker will backtrack and carry out a sweep from the last point of confirmed sign. This sweep will generally take the form of walking in a circle some metres from the last known sign.

EVASION PLOYS

The quarry may try various ploys to throw the tracker off the trail. However, most such efforts, unless properly executed, will be in vain and only confirm that the quarry is changing direction. Any delaying tactics the quarry may use must buy more time than that expended on the delaying manoeuvre. The most commonly used ploys are:

Walking backwards When a person walks backwards the length of his stride is shortened. The imprint of the toe and ball of the foot will be more pronounced. Loose dirt, sand or leaves will be dragged in the direction of the move. Always walk backwards on your heel, lifting your knees slightly in order to imitate the action of walking forwards.

Brushing the track Trying to sweep the track with branches only signposts to the tracker the quarry's intention to change

When done convincingly, the simple trick of walking backwards can sometimes fool a tracker into heading in the wrong direction.

direction. It is better to walk along the side of the track for some distance, as this will reduce the spoor to nothing.

Crawling The sudden absence of any top sign will indicate that the quarry is crawling or is injured. Crawling is a good ploy to use if you come across the trail of a large animal, as any signs you leave may be confused with those left by the animal.

Booby-traps and man-traps Unless the quarry has access to military mines and equipment, the time needed to construct a booby-trap or man-trap is time wasted unless it is guaranteed to delay the tracker. If you do intend to build a trap, try laying three or four well-concealed fake trip-wires before you lay the real thing. This will unsettle the tracker, making him slower and more cautious.

Warning: Information on all traps in this book is given exclusively in the context of evasion or collecting animal food. Neither the author nor the publisher accepts responsibility for the results of their use in other contexts.

Speed If you are fit, sheer speed may well put enough distance between yourself and the tracker to allow the spoor to become cold. This is best achieved where the area is open and the ground sign you leave is light. If you intend to benefit from speed, try it in the early days of your escape, when your energy and fitness level should still be high.

Irrational actions A useful evasive ploy is to combine speed with feigned irrational behaviour. Doubling back parallel to the line of your march for some distance will confuse the tracker. Climbing a rock-face when it is not necessary or building a camp-fire out in the open should have a similar effect. Remember that the tracker

Attempting to brush away boot prints with a branch seldom puts an experienced tracker off the trail.

Crawling creates a clear sign for the tracker. However, the evader can use it to simulate the passage of an animal.

Man-traps take precious time to make, so it is far better to use irrational actions to throw the tracker off your trail.

Broad rivers are great escape lines, especially if they flow in the right direction for your purposes. Use them at night and drift with the current.

will not be alone, and if you can sow seeds of doubt about his abilities in the minds of others, it will help to frustrate attempts to capture you.

Leaving a clear false trail will also work but don't overdo it. Walking along railway lines for some distance and then jumping off the track will grant you some time. A large river can provide a simple, effortless mode of transport. But stealing a boat will bring attention to your method of escape, so it is safer to build a raft of drift-wood, and best to move at night.

4

SURVIVAL

The shock of finding yourself isolated behind enemy lines, and the possibility of being captured, will serve to heighten the urgency you feel to survive and stay free. But you should beware of thinking of yourself as a rootless vagrant. Instead, concentrate on acclimatizing yourself to your surroundings so that you feel like a native. Doing this will help to restore an all-important feeling of pride in yourself that will help you to get through a difficult situation.

At the same time as using all your training in survival to avoid visual exposure and the possibility of capture, you will have to meet the requirements of day-to-day living. If you keep clean, do your best to find water and food so as to keep your strength up, rest adequately and navigate accurately, you will greatly increase your chances of returning before long to a friendly environment. To aid you in this goal, there are a number of tried-and-tested practices and items of equipment, and these are the subject of this chapter.

SURVIVAL ASSESSMENT AND PLANNING

The first reaction to any dangerous emergency will almost certainly be instinctive, for self-preservation is the strongest drive we have. The alarm and fear that danger engenders makes the adrenalin flow, stimulating the body into a state of alertness. But as soon as the immediate personal threat has passed, logic and careful observation take over in those who have been trained to deal with emergencies. The overwhelming requirement in a survival situation, and indeed in any kind of emergency, is to think clearly. Your finest asset is your brain and you must begin using it as soon as possible in order to evaluate the situation and plan activity that will contribute to your survival.

If you are not alone, the first thing to do is to check carefully if anyone else needs medical help, and to administer first aid if it is required. When you are sure that no one is in immediate danger you should carry out the following actions:

1 Put on any appropriate clothing available, to conserve body heat and/or body fluids. The choice of clothing depends on the weather conditions. Covering the head is just as important as covering the body, as some 30 per cent of body heat can be lost through an uncovered head. If you have no purpose-made headgear, improvise, using clothing, for example. Keep as dry as possible, because when it is wet clothing can lose up to 90 per cent of its insulating properties.

2 Gather together any survival equipment you have brought with you, all the available food and water, and any items that might prove useful. Interpret the word 'useful' as imaginatively as possible, for the most unlikely object may provide the solution to a particular problem. For example, an electrical generator may not appear to be much help in providing food, but it contains wire which will make effective snares and serve other purposes besides.

3 Assess your situation as it really is. Put aside any thoughts of the 'if only' kind and instead maintain a strictly realistic approach. Take this opportunity to relax – this will help you to think more clearly and make you feel better physically – and in doing so try to be more receptive to what you see, hear, smell and feel. Being attuned to your surroundings in this way will help you in practical ways; most importantly, to sense danger and avoid capture.

4 Decide whether shelter will be needed against cold, wind, rain or snow. If the temperature is low, remember that the most dangerous threat will come from the cold, which can kill a person much more rapidly than lack of food or water. Seek a shelter, even if it is to be used only temporarily, making full use of

When you are in pain, understanding its source will help you to bear it. Focus your mind on what to do next, and concentrate on surviving.

Extreme cold numbs the mind and dulls the senses. In a cold environment, take every opportunity to create extra warmth. Clothing should be loose and it must be kept clean, as dirt causes wear. Protect the head from heat loss at all times.

Fatigue is a dangerous condition for a survivor. To minimize the risk of falling prey to it, seek shelter from hot, cold and wet weather. Weigh up the energy expenditure of any actions you contemplate against their likely worth in helping you to survive.

Look for natural shelter such as caves or hollows in the ground. Then build a fire, so that you stay warm, dry and healthy. This done, give yourself time to think about your further survival plans.

any natural features such as caves, the lee side of rocks, trees, fallen logs or snow banks. Add a covering to keep off the rain as soon as possible, but don't waste time and energy building a shelter if nature already provides a usable one. If you decide that a shelter needs to be constructed from scratch, you will find that the most serviceable materials are branches from trees and anything that can be used to build a tent-like structure or a lean-to.

A fire with which to keep warm and cook may be another need you will experience early on, but you should make a fire only if there seems to be no risk of the smoke giving away your position. Check that you are sufficiently far from pursuers or the enemy in general for a fire not to prove a danger.

When these steps have been taken, it is time to formulate a plan for future action. This should be based on a set of priorities which will become clearer once you have answered these questions:

- Are there any other survivors separated from you or your group?
- If you or anyone else is injured, are you or they capable of moving or being moved?
- What are the weather prospects?
- Would it be possible to move through the surrounding terrain in safety?
- What is your present position?
- In which direction, and how far away, does the nearest help lie?
- What stocks of food and water do you have to hand?
- What are the prospects of supplementing your food and water from the countryside around you?
- What equipment do you have? (e.g. survival or first-aid kits, tools etc.)
- Do you have communications – e.g. radio, tacbe or distress beacons?
- Are you being hunted?

Once you have made decisions based on your answers to these questions, stick to them unless some major new circumstance arises which is important enough to warrant a change of plan.

Probably the earliest and most vital decision will be the choice between staying where you are and awaiting rescue, and attempting to make your way to safety. There is no simple answer. The circumstances of your survival situation – and there are many variations – will, to a greater or lesser extent, influence the choice you make. For example, you may have become lost in

After an aircraft or vehicle crashes, staying put may be the best solution. If it is safe to do so, use the wreckage as shelter. Make sure any fire you make is well away from fuel. If you intend to move, salvage items that may be of use, such as medical kits, water, maps, and compasses. However, carry only what you really need. Share the load between survivors, according to age, size and fitness.

the mountains and/or be threatened by the onset of foul weather. You may be a survivor of an aircraft which has crashed, or had to make an emergency landing, in a remote area. Alternatively, you may be a survivor of a shipwreck, and either cast ashore or drifting in a small vessel.

If you have escaped from captivity, a crucial question is how long it will be before you are missed and a search party is organized – by your comrades or the enemy, or both. If your emergency has arisen because your deep-penetration patrol has vehicle problems, or your aircraft has been shot down, bear in mind that a search-and-rescue team is more likely to spot wreckage – of an aircraft, ship or motor vehicle – than individuals. In addition, if you decide to stay and await rescue, you can do a number of things to make your location more visible. You will also be staying close to whatever resources are offered by the vehicle or its wreckage. However, take into consideration the fact that the enemy will also be looking for you.

If it is safe to do so, check the surrounding area thoroughly, but without wandering very far away. By doing this you will probably be able to make a better assessment of the ease with which you can move about in the locality, and you should also be able to establish the availability of water, fuel and food. Having taken time and trouble to collect the best information obtainable, be sure to make good use of it. Always remember that you are looking for the safest option – the choice that will give you the best chance of survival.

If you have good reason to believe that rescue is imminent, and have decided to stay put, ask yourself several further questions. Where will the rescuers be coming from? Where are they likely to start searching? How are they going to come? What signs will they be looking for? How can you make these signs more obvious? If you are injured and alone you may not be able to stand to make your presence known to rescuers arriving in your vicinity, so how can you alert them?

If you have decided to move, consider carefully what you will be able to take with you. Your choice may have to be a compromise between what is available and how much can be conveniently carried. If you are in a group, make sure that all available protective clothing

With rescue imminent, signal your location as best you can.

has been shared equally and that everyone is as well protected as possible.

Select what you are taking with great care. Naturally you should regard any survival supplies – first-aid items, food and water, survival items, signalling equipment such as flares, and so on – as a priority, selecting these first and then adding to them anything which may be useful but is not bulky or weighty. Use the same criteria as you would for assembling your own personal survival kit – but with ingenuity and an eye for versatility and adaptability. While you are on the move you will need to find or construct shelter in different places, as well as obtaining warmth, food and water – all of which are important if you make overnight stops.

CLOTHING AND HYGIENE

Man is essentially a tropical animal, and in most parts of the world he needs clothes to protect himself against the elements. In a survival situation maintaining body temperature, along with avoiding injury, is just as important as finding food and water. The body functions best at temperatures between 35.5°C (96°F) and 39°C (102°F), and therefore the prevention of heat loss or gain that takes body temperature beyond these limits is of primary concern to the survivor. Factors which cause changes in body temperature are climatic temperature, wind, moisture loss and illness. The process of heat loss or gain occurs through conduction, convection, radiation, evaporation and respiration.

EFFICIENT CLOTHING
Wearing the right clothing for the expected conditions is of great importance to the soldier, as it is to anyone in a survival situation and indeed someone simply requiring protection from the elements during a leisure pursuit. Although the modern soldier in most armies is equipped with state-

> **Survival tip**
> The wind-chill factor is the greatest threat to any survivor. It can rob the body of heat in cold-wet conditions and of moisture in hot conditions. Do everything in your power to prevent it. The primary protection is appropriate clothing, properly fastened to prevent exposure to cold air. But the use of shelter from extreme wind-chill conditions plays an essential role in maintaining body temperature.

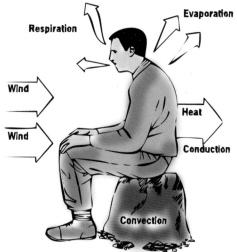

The body loses valuable heat in several ways.

of-the-art protective clothing, in survival situations various factors may reduce its benefits or remove them altogether. For example, long-term survival will take its toll on clothing and equipment, reducing its efficiency, while enemy troops who are not so well equipped may relieve a captured soldier of his superior uniform, weapons and other equipment.

The principle of using a layer system in order to trap warm air around the body has proved to offer the most benefit in cold conditions. The combination of thin cotton underclothes with a covering of one or more layers of warm clothing made from wool, fibre pile or fleece, is ideal. On top of these is worn an outer garment, which should be windproof and preferably waterproof. It can be made from tightly woven cotton, polycotton, fibre-pile material or nylon. In order to derive the maximum protection from clothing, act on the following points:

1 Keep clothes clean
2 Avoid overheating
3 Keep clothes dry
4 Repair defects immediately
5 Improvise.

Any item of clothing with more than one layer will trap warm air around the part of the body or other clothing that it covers. If you wear two pairs of gloves, be sure that the outer pair are large enough to allow this to happen.

CLEANING CLOTHES

By keeping your clothes clean you help to prevent the build-up of grit that destroys the fibres. A garment weakened in this way is less likely to survive the strains it is subject to and if it tears or is holed it will be less effective in keeping out the cold, rain and snow. However, in situations where washing is not practical a good daily shake or beating will suffice. Items of clothing that lie next to the skin, particularly socks and underclothes, require daily attention, whether it be washing or beating. Take care when beating modern army clothing against rocks in the traditional manner, since buttons and zips can be damaged, particularly the latter, which you will almost certainly be unable to replace in a survival situation. In tropical climates clothes need to be washed more frequently than in cooler conditions, to prevent them rotting from the combination of heat and sweat.

OVERHEATING

Clothing should not be worn so tight that it restricts the flow of blood, since it is the circulatory system that distributes body heat and in cold conditions helps to prevent frostbite. If you are wearing more than one pair of socks or gloves, make sure that the outer pair is large enough to fit comfortably over the inner pair. Loosen any clothing at the neck, wrists and waist during excessive exercise. If the body is still over-heating, experiment by taking off one layer at a time. When work stops you will be hot and sweaty, and you should put the clothing on again immediately to prevent yourself catching a chill.

DRYING CLOTHES

Overheating results in perspiration, which wets clothing and reduces its insulation qualities. Ventilation is essential when physical exertion is unavoidable. Clothing that is wet from sweat – or rain – will lose heat up to twenty-five times more quickly than dry clothing. A combination of wet clothing and strong winds can lead to a swift death.

Wet clothes can be dried in various ways. Lay them out on clean rocks that have been warmed by the sun. In windy weather, hang them securely from branches. If tactical conditions allow, build a fire, but never leave clothing drying in this way unattended, as there is a risk of burning this vital part of your survival equipment. Take care when drying leather boots or gloves by a fire – if they dry too fast they will stiffen and crack.

In sub-zero temperatures clothing can be dried by hanging it up to freeze and then beating it to remove the ice particles. This works best on tightly woven garments. If you should be unfortunate enough to fall into water, try rolling in powdery snow as it will soak up the moisture to some extent.

The hide from various large animals can be used to make effective footwear. To do this:
1. Remove the limb totally.
2. Strip off the outer skin as an intact tube, cut it to the required length and make lace holes.
3. Sew up the toe end of the tube. Laces can be made from strips of dried hide.

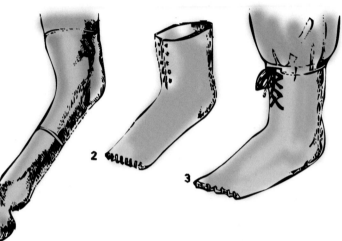

WATERPROOF SOCKS

Waterproof and breathable socks Made from Aquatex with fully taped seams, over-sock are available that are both 'breathable' and water-proof. They are designed to fit over the normal inner sock, keeping the feet warm and dry even in the worst weather conditions.

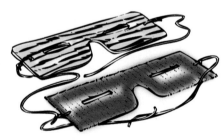

Protect your eyes from the snow's glare with sunshades improvised from card, wood, plastic or even grass.

REPAIRING CLOTHES

Adopt the Eskimo habit of repairing clothing the moment it is damaged. Pay particular attention to items that are windproof. Never modify any of your clothing for the sake of comfort. For example, in hot climates do not cut off your trousers to make shorts, as you may later need the covering for your lower legs. This is particularly important when you are in the desert, where, although the heat is intense during the day, the temperature falls dramatically at night, often to below freezing point.

IMPROVISED CLOTHING

There are various means by which a survivor can supplement his normal clothing and equipment in order to protect his body.

Sunshades Snow-blindness is a severe problem, and if looking at the snow starts to hurt your eyes make a set of improvised goggles. A variety of readily available materials can be used for this purpose.

Improvised boots Animal hides and bark from suitable species of tree can be used to make crude but serviceable boots that can be used in a variety of climates and local conditions.

SLEEPING COMFORTABLY

As a survivor you will spend at least one third of your time resting or sleeping. It is vital to make this time comfortable since doing so will help both your mental and physical well-being. If you are lucky enough to have a sleeping bag, make sure that you turn it inside out each morning and give it a good shake before storing it. In the evening, 'fluff' it out to give maximum insulation. Never get into a sleeping bag in wet clothes, as this is likely to lead to a chill.

A sleeping bag can be made from a simple polythene sack. This will help to protect you from wind and rain, but better is a hessian sack inside a plastic sack, which will prove warmer and more comfortable. Insulation can be improved by filling the gap between the inner and outer sack with insulating material such as straw, grass, moss or cardboard. I proved the efficiency of this ploy in Arctic Canada, on snow and in temperatures of -10° (14°F).

Russian socks are a tried-and-tested means of keeping your feet warm whether you have boots or not. Each foot is placed on three triangles of cloth, above, and wrapped in it. The cloth is then tied at the ankle, above right.

Russian socks The footwear known as Russian socks can be made from triangular pieces of a material and used as an inner sock or complete foot covering. You can use any supple material, but the stronger the better.

1 Cut out three 1-metre (39in) squares from the material. Lay them on top of each other and fold them

Gaiters support the ankles, reducing the risk of injury.

A head net is essential to give protection against mosquitoes and other insects that bite or transmit disease. Make one from any suitable light fabric.

once to form a triangle.

2 Place the foot in the centre of the triangle of cloth, with the heel facing the fold and the toes facing towards the corner.

3 Fold the corner up over the toes and wrap the two side corners over the instep.

4 Repeat the above procedure with the other foot and slip your boots over your swaddled feet. If you have no boots, fashion an insole from a piece of thick cardboard and place it under your foot before you wrap the Russian sock around it. Finally, secure the sock with string or cord.

Gaiters You can give support to the ankle by using the time-honoured device of gaiters. In the absence of military-style gaiters, these can be improvised from strips of cloth of about 1 metre (40in) in length by 12cm (5in) in width.

Insect nets Mosquitoes and certain other insects can cause severe problems, not just by inflicting a painful bite or sting but by transmitting a disease to the victim. Wearing proper clothing and ensuring that no bare skin is exposed will reduce the risk of this occurring. Fashion a reliable head net – it should be sufficiently large not to touch the skin of the face or neck – and in areas of high insect infestation you should always protect your hands with gloves.

PERSONAL HYGIENE

Maintaining personal hygiene in trying or dangerous conditions is unlikely to be as difficult for the soldier as it will be for the civilian who finds himself in a survival situation, since the former's training will have prepared him appropriately. In the context of survival, personal hygiene embraces not just bodily cleanliness but also protection against disease.

WASHING
Bodily cleanliness is a major protection against germs that carry disease and against infestations of ticks, lice

and fleas. A daily all-over wash with warm water and soap is the ideal. But if this is not possible, concentrate on keeping your hands clean and wash and sponge your face, armpits, crotch and feet at least once a day. In desert conditions scrubbing the hands and feet with sand will suffice, as will using washing with snow in cold conditions. In the absence of any alternative, take an 'air-bath', rubbing the body briskly with a clean piece of cloth. Clothing, especially underclothing, must be kept as clean and dry as possible – for health reasons as well as to prevent wear. At the very least, you should shake it out and expose it to the sun and air each day. Also, check for and remove parasitic insects regularly.

Wash your body as often as possible to increase protection against disease.

You can protect your teeth by improvising a toothbrush or toothpick from the crushed end of a small twig. Charcoal ash is an effective tooth powder.

FEET AND FOOTWEAR

Keeping your feet and boots in good condition is one of the basic lessons of survival, since no one forced to travel on foot will get very far without taking great care of feet and footwear alike. In fact many people who were otherwise fit have died because of lack of foot care and protection. *Never forget: your feet may be your only form of transport.*

A stick, crushed at one end, will serve as an efficient toothbrush, while ground charcoal will help to clean the teeth.

All new footwear must be broken in before you attempt any prolonged walking. If it is cold and your boots are big enough, use straw, grass, cardboard or any other dry insulating material to maintain warmth in your feet. In snowy conditions warmth in the feet can be increased by wearing large socks over boots.

However good your boots might be, blisters on the feet are inevitable when you do a great deal of walking. If possible, you should remove all blisters carefully. Always avoid breaking them to reveal an open sore, and instead prick the edge with a sterilized needle. Blisters will heal more quickly if they are kept clean and dry. Bathing the feet in hot, salt water soothes them very effectively and helps to harden the skin against the formation of further blisters.

A SURVIVAL KIT

Pilots and some special forces are issued with escape and survival kits, but with a little thought anyone can assemble a basic but effective kit. With a properly planned survival kit you will possess the resources to make full use of the survival techniques you have learned. Your personal kit is extremely important in a survival situation as it is the catalyst which stimulates constructive thought and spurs you on to start planning a practical approach to your survival.

The range of survival equipment is vast, with new items being added all the time. The list below is a basic one, based on practical experience and with no fancy extras. You may well be forced to compromise in your choice – between what you feel you need, and the amount you can comfortably carry. You may well consider items other than those listed below which you feel should be in your personal kit. But ask the following questions about each one. Is it really necessary? Is its function duplicated by any other item? Remember that the aim is to keep your kit as small as possible.

You must assess every item's usefulness, its adaptability and its weight or bulk. Make this assessment while keeping in mind the strong possibility that the kit may be your only initial resource. Select items on the twin criteria that they are small enough for you to carry at all relevant times, and that each will increase your chances of surviving, or escaping. Remember too that these items will be the tools with which your survival skills can be unlocked and exploited.

Another factor to consider is that the choice of contents in your survival kit should reflect the type of operation in which you will be engaged and the location in

A survival kit intended for use in the jungle must contain medical supplies suited to that environment, in addition to items found in a standard survival kit. An essential item is a supply of insect repellent.

In compiling an arctic survival kit, emphasize the ability to make fire and create shelter.

which it is to take place. In particular, different environments and climates will make different demands on your kit. It is worth thinking about substitutes for items in your present kit. If necessary, these reserve items can be used to modify it before you set off.

The personal survival kit should be carried – always on your person – whenever there is the risk of finding yourself in a life-threatening or otherwise hazardous situation. However, if the possibility of capture is

Most items included in a survival kit are simple, yet they can make the difference between life and death. Select every item with great care since space will be precious and weight must be saved, giving thought to the nature of the terrain you will be in.

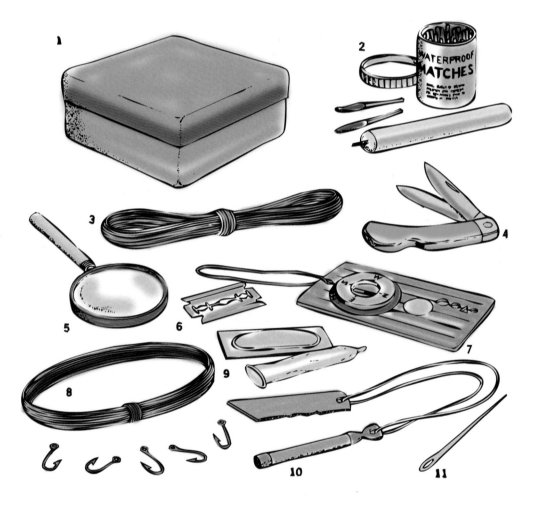

1 Watertight container
2 Waterproof matches and candle
3 Brass wire
4 Pocket knife
5 Burning glass
6 Razor blade
7 Compass
8 Fishing line and hooks
9 Condom
10 Flint and steel
11 Needle

Protect matches from damp by coating them with wax.

Every survival kit should contain survival matches.

imminent, do your best to hide the most useful items in your clothing and on your body and abandon the rest.

MATCHES

A dozen or more household matches which have been completely immersed in melted candle wax will be both waterproof and wind-resistant. They should be carried in a waterproof container, and packed loosely so that one can be withdrawn easily even when the fingers are cold, wet or numb. A scrap of emery cloth will provide a dry striking surface, but it must be stored in the waterproof container in such a way that it will not make contact with the match heads.

Survival matches are available which are either windproof or waterproof or both. These can be bought from camping and outdoor pursuit shops, and are included in some ration packs. These matches, of better quality than the domestic variety – each is hand-made and coated with protective varnish – are sold packed in airtight tubes. They burn for about twelve seconds, and will not go out even if they are completely immersed in water or exposed to the strongest wind.

FIRE-LIGHTERS

A simple fire-lighter consisting of a flint and a steel striker is a rugged and dependable item of equipment which will light countless fires in all kinds of weather. Carry it as part of a 'fire set' which also incorporates other essentials for lighting fires in hostile conditions.

A good fire set will be equipped with a flint containing magnesium, to produce large, very hot sparks. The striker will have hardened teeth to ensure effective action against the flint. It is a good idea to sharpen its leading edge to provide a small, sharp blade. This could be useful in many ways, not least for scraping fire tinder from clothing.

A small supply of cotton wool, which makes excellent tinder, will be another component of the fire set. Also included will be fire-making blocks, which are a fine primary fuel. They burn slowly in block form but provide rapid ignition when crumbled. Incidentally, the tin in which these blocks are purchased has a variety of possible uses – for example, as a heliograph or to hold animal fat, which can be burned to serve as a simple lamp.

TAMPON

Skill in lighting a fire depends on the quality of the spark supply but also on that of the tinder. Over the years I have experimented with many types of tinder and to date by far the best I have come across is the cotton wool in a tampon. A tampon for this purpose is now standard issue in most RAF pilot survival kits. If possible, blacken the cotton wool first with old char coal, as this will make it accept the spark more readily.

A flint and a steel striker produce sparks that will ignite cotton wool or other tinder.

CANDLE

A 10cm (4in) candle weighs less than 30gm (1oz) yet will burn for up to three hours if it is protected from the wind. The best choice is a candle made from 100 per cent stearin (solidified edible animal fat). This will light in any temperature, serve as food in an emergency and lubricate items as diverse as a zipper and the hand-held socket of a bow-drill fire-maker. It is impervious to dropping and soaking in water, and practically indestructible. In addition to providing light, a candle will, when burned in a tin can, heat a small, snug shelter, snow cave or igloo. When other fuel is not available, it will ignite tinder too damp to be lit by matches.

A candle in a tin can provides light and, when used in a small shelter, warmth.

CONDOM

The condom has one of the widest ranges of uses of all the items in a personal survival kit. When held in a sock or shirtsleeve it becomes a water carrier. Note, however, that a condom cannot be filled by simply dipping it into a water supply. To exploit its storage capacity adequately, you must pour the water in slowly. If a condom is carried as described above, it will hold approximately 1.5 litres (2.6 pints) when extended to a

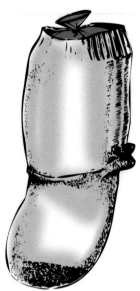

A condom in a sock makes a useful water carrier.

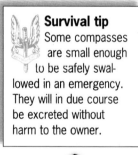

Survival tip
Some compasses are small enough to be safely swallowed in an emergency. They will in due course be excreted without harm to the owner.

A Silver compass, above, is a fine tool but a button compass, right, takes up less room.

length of 30cm (12in). A condom can also be used as a waterproof container for small and medium-sized items which require protection from damp (e.g. dry fire tinder) Two condoms together can be used as a catapult. For all these uses the condoms chosen should be the inexpensive variety, non-lubricated and heavy-duty. Your kit should include a minimum of three.

NEEDLE

A needle with a large eye of about 5cm (2in) in length – for example, a Chenille No. 16 or a sailmaker's needle – can be used for heavy-duty sewing of shoe leather, hide or heavy fabric. It can also be magnetized and used as a pointer in an improvised compass.

COMPASS

A compass is an invaluable item in any survival kit. A button compass is the ideal choice – easy to read but occupying the minimum of space. The 'Explorer' button compass, filled with liquid, is a good choice. This sort of compass should be fitted with some form of lanyard to guard against loss – a real possibility because of its small size. Several good alternative patterns of compass are available, including the Silver, but most will take up considerably more precious room. Whatever kind of compass you choose, it is very important that it should be a good-quality instrument.

SURVIVAL BAG

One of the most frequent dangers to be faced in a survival situation is the involuntary loss of a critical proportion of body heat. A green or camouflage polythene survival bag is the best choice for your personal survival kit. It counteracts most efficiently heat loss that occurs through convection. Once you are inside the bag, your body will be protected from wind and rain. Heat loss which occurs as a result of conduction can be minimized by laying the bag on a bed of an insulating material such as bracken, straw or grass. A survival bag can also play a role in carrying water, and as part of a shelter from the elements.

Saw

Including a wire saw in your kit will allow you to cut down trees or branches – to provide covering for yourself, for use in making a shelter, and for firewood – and the better examples will cut through materials such as bone, plastic and iron. A saw needs to be made from at least eight strands of wire if it is to have the strength it requires for lasting use. Nevertheless, wire saws can be broken fairly easily. Always saw slowly, so that the wire does not overheat. After use, the saw should be held straight and taut until it cools.

For long-term or one-handed use, make a bow-saw using a suitable green stick. Used with rope or cord, the saw can cut overhead branches from ground level, so as to make a clearing in wooded areas or bring down fruit-laden branches. The smaller ring fitted at one end

It is important to retain body heat when you are asleep, and for this it is best to use a purpose-made survival bag. Where possible, improve heat retention by placing the bag on bracken, straw or grass.

of the saw will pass through the larger ring at the other end, allowing the saw to become an efficient snare.

Water-purification tablets

Sterilizing water is frequently a necessity in a survival situation, and water-purification tablets provide a

A wire saw, above left, will cut off tree branches for firewood or building a shelter, but its use calls for two people. One person can use a bow-saw, above, made from a green stick and rope or cord.

quick and convenient way of doing this. Each tablet will purify 1 litre (1.8 pints) of water in about ten minutes. Although the tablets will kill bacteria, they do not remove any dirt present in the water, and the treated water tastes of chlorine. It is a good idea to carry about fifty tablets in your kit.

KNIFE

The Swiss Army knife, which includes among its many implements several blades of different sizes, scissors, can and bottle openers, a screwdriver and a tiny saw, is an ideal choice for your kit. It is strongly recommended that you carry a small pocket knife of this kind as a matter of course.

However, there is an alternative to the long-established pocket knife – the purpose-designed survival knife. During recent years sales of survival knives have increased dramatically. The number of facilities they offer varies between knives, but in general the better-quality models a more versatile. Unfortunately cheap, poorly made survival knives, mainly from the Far East, have flooded on to the Western market, and great care must be taken to avoid these when you are making your choice. You cannot afford to have an item of equipment on which your life may depend, fail you at a critical moment.

The Leatherman is an excellent all-round tool, its many applications making it one of the best of all survival items. Made from stainless steel which has been especially hardened to last a lifetime, it includes: full-size pliers, wire-cutters, knife blade, can and bottle opener, four types of screwdriver, and a file suitable for metal and wood.

SNARE

It is possible to make four snares if your survival kit contains, as is advisable, 3 metres (10 feet) of brass snare wire. However, it is better to include ready-made snares in addition to the wire, with the latter serving as a stand-by. Ready-made snares are cheap, and work far

A sturdy multi-bladed pocket knife is an invaluable part of any survival kit.

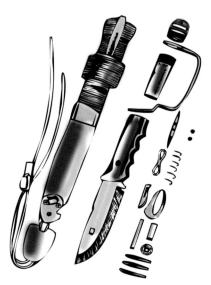

A large single-bladed knife, used for cutting meat and sizeable items, complements the folding knife in a kit.

more effectively than wire. In either case, the snares should be rubbed with animal droppings before they are set. As well as trapping animals, they can also be used to catch waterfowl and fish.

RAZOR BLADES

In most military survival kits razor blades with one reinforced edge for gripping are standard issue. They will serve a wide variety of cutting purposes, from making weapons to skinning game and gutting fish. Always treat a blade with care and limit the type of materials you cut to its capability.

MAGNIFYING GLASS

It is not easy to light a fire with a magnifying glass. The tinder must be very dry, and it will require a great deal of gentle blowing to ignite the smouldering sun spot so that a flame occurs. The glass can also be used for locating irritating splinters and thorns

PARACHUTE CORD

Many a farmer will tell you that he never goes out without a quantity of string in his pocket, so versatile is this simple item. The same principle should apply to survivors, although you will be better off using para-chute cord (usually referred to as 'para cord'). This extremely strong cord, with a breaking strain of about 250kg (550lb), is braided over strands of thinner cord, which can be pulled out and used for thread or fishing. As parachute cord is so versatile, you should carry at least 15 metres (50 feet) in your kit.

HELIOGRAPH

There are various types of heliograph, or signalling mirror, on the market, most of them requiring two-handed operation. In the past few years, however, a new one-handed model has entered the market and is very effective. The accuracy of these new heliographs is very high, but you should take time to practise using one before venturing into the wild.

> **Survival tip**
> It is important that the survival kit is not opened until it is needed. The only exception to this rule is to modify its contents because the operation or journey in prospect will be taking place in an environment different from that originally envisaged. For example, if you are going into a desert area you should include more items that will enable you to collect water.

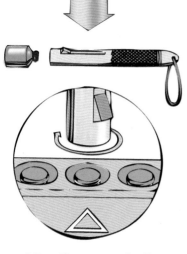

A flare kit, top, contains flares of up to nine different colours and a pistol, centre, for launching them. Once a flare has been screwed into the pistol, above, the loaded pistol must be handled with great care and held facing skywards before it is fired.

FLARES

In a survival situation, signal flares attract attention better than most other methods. A standard flare kit protectively houses up to nine different-coloured flares and a launch pistol. Great care should be taken when firing the flares, and in particular you should always point the flare skywards before firing.

FISHING TACKLE

A small survival fishing kit can be made up from about 30 metres (100 feet) of nylon line with a breaking strain of around 2kg (4.4lb), five hooks (barbed, size 14 or 16) and swivels, and ten weights (iron or brass). A brewer's cork will serve as a float. The cork, if charred, will provide face and hand camouflage - helpful if you are hunting. A plastic luminous lure is also useful. Worms, slugs, caterpillars and other insects make good baits, as does bread, if you can spare any.

SURVIVAL MEDICAL KIT

You should also assemble a small personal medical kit. The items chosen should reflect your skill in first-aid in the field, but should include the basic following items:

Plasters and dressings Provide a good selection of assorted plasters, but include more large than small ones as they can be cut to size according to your requirements. Choose the waterproof type. A strip of butterfly sutures is ideal for closing small wounds.

Always carry at least one large wound dressing. In all my military career I was never without one, and had several occasions had desperate need of it. (Never discard a used wound dressing, as the cotton wool inside it is a valuable source of tinder.)

Aspirin Include a strip of a dozen soluble aspirin for relieving mild pain and headaches. Aspirin can also reduce fever.

Mosquito repellent If you plan to venture into areas infested with mosquitoes, take an effective insect repellent. In this situation anti-malaria tablets should be consumed as a matter of course.

Antihistamine cream The severe irritation which can result from insect bites and stings can be relieved with antihistamine cream, so it is essential to pack a tube of this medication.

Antiseptic When it is mixed with water, potassium permanganate makes an excellent antiseptic that can be used for cleaning wounds or sterilization. Dry, it can be used for starting a fire.

Salt A small waterproof container of salt should be carried if you are travelling to tropical climates. It is best reserved for medical purposes rather than used for flavouring food.

Medical items are among the most important elements in a survival kit. In a hostile environment it is vital to remain as healthy and as unimpeded by injury as possible. Various items, such as plasters and dressings, aspirin and antiseptic, should form part of any survival kit, but others will be of more use in particular conditions than in others. For example, a good supply of insect repellent will be needed in climates where troublesome insects abound.

PACKING SURVIVAL ITEMS

Having subjected all the items on your checklist to rigorous scrutiny, assemble all those which have passed the test. Make sure they are well packed and totally waterproof. The simplest way of ensuring the latter is to seal the whole kit in an airtight container. A metal tobacco tin is ideal. Alternatives include a waterproof plastic box, a resealable polythene bag inside a heavy-duty canvas pouch, a snap-seal plastic container (of the type used to pack tablets) or a screw-top cylindrical metal container. Once the survival kit container has been packed, it should be sealed securely with adhesive polythene tape.

Navigation will also be a major priority, as will signalling, since there is very little to burn in the desert to indicate your presence. Therefore you will need to pack equipment to meet these needs. A journey to a cold region calls for items to provide warmth, shelter and possibly a spare fishing kit. Travel to a jungle or other hot area demands extra insect repellent and medical supplies to deal with bites and stings, and problems caused by the heat.

IMMUNIZATION

All military personnel and others intending to enter environments where there is a high risk of contracting a disease, should make sure that they are immunized against as many diseases as possible, or, if they have received immunization some time earlier, that this is still effective. Typhoid, paratyphoid, yellow fever, typhus, tetanus, smallpox and cholera are among the diseases that can be prevented by immunization. Seek medical advice on any special precautions that are required for the areas you intend to go to.

5

FIRST AID
FOR SURVIVAL

The ability to carry out first aid promptly and efficiently is of great value in normal circumstances. In a survival situation, where there may be no prospect of skilled assistance, the value of this expertise is beyond measure. Even though medical supplies in such a situation may well be inadequate or entirely absent, you may have to provide first aid for yourself or other survivors. Restricted though the possibilities for treatment may be, the combination of even limited first-aid skills knowledge and improvised equipment can save lives.

PRIORITIES OF FIRST AID

Below I describe only first-aid practices that are applicable, and have been tried and tested, in extreme survival conditions, such as experienced by a captured POW or an escapee. As in every other aspect of a survival situation, the need for first aid must be assessed, the priorities established, and a course of action planned and carried out. The situation itself will largely determine what is decided, but whatever the circumstances, bear these general rules in mind:

- Keep calm. However serious an injury or dangerous a situation, panic will impair your ability to think and so lower your effectiveness. Also, time will be wasted – and in a crisis time can mean the difference between survival and death.
- Avoid any unnecessary danger to yourself. This is not cowardice. You will be of no help to anybody if you suffer avoidable injury.
- Think carefully and calmly, but also as quickly as possible, before you act.
- Do your best to reassure and comfort any casualty.
- Find out if there are any other uninjured or active survivors who can help you to deal with the situation. In particular, seek out any survivor who has

medical qualifications or experience that is better than yours.

● In assessing individual casualties, use your own senses to the full. Ask. Look. Listen. Smell. Then think and act. Ask the casualty to describe his symptoms, and to tell you what he thinks happened and what he feels is wrong with him. Then check the following.

BREATHING

To determine if an unconscious casualty is breathing, listen with your ear close to his nose and mouth. You should be able to hear and feel any breath. At the same time watch out for chest and abdominal movement. If there is no sign of breathing, then without delay take the following action to ensure that the air passages are clear.

(a) Supporting the neck with one hand, ease the head backwards with the other. Keeping the head back, lift the chin upwards. This will open the air passage and bring the tongue forward to prevent it from obstructing the intake of air. Check quickly inside the mouth to find and remove any other cause of blockage – for example, dentures, vomit or other materials. Once the air passage is open and clear, the casualty may begin breathing again. If this happens, and his heart is beating, put him into the coma (or recovery) position (*see* page 82). If there is a visible injury to the front or back of the head (the latter may indicate damage to the neck or spine), maintain a clear airway by keeping the head back. Improvise some form of collar or head support to keep the head correctly positioned for breathing.

(b) If breathing does not recommence, the casualty must be given help with respiration. This can best be done on a mouth-to-mouth basis. Taking a deep breath, pinch the casualty's

Listen and smell for breath from the casualty. Lay the head back and open the mouth to allow the passage of air. Remove false teeth and vomit or other debris.

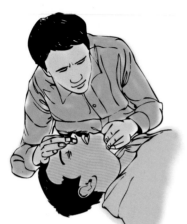

1 Pinch nose and blow into mouth to restore breathing.

2 During resuscitation, check if chest is rising and falling.

3 Check the right side of the neck for the carotid pulse.

nose to prevent air loss, open your own mouth wide and seal your own lips around his open mouth. Blow into his lungs, watching for expansion of the chest. When the maximum expansion is reached, raise your head well clear and breathe out and in again. Look for contraction of the chest. When this has happened, repeat the procedure four times. It may be more convenient to use mouth-to-nose contact. In this case, the casualty's mouth must be kept shut to prevent the loss of air.

After the fourth assisted breath, it is important to check that the casualty's heart is beating. Oxygen, having been taken up by the blood, must be delivered to the body's vital organs. Feel for the carotid pulse, in the right side of the neck. If there is no heartbeat, chest compression must be carried out as described below. Be sure that there is no heartbeat before starting to carry out chest compression. Far more harm than good will be done if attempted chest compression interferes with an existing heartbeat, however weak the latter.

If the heart is beating, continue to give assisted breaths at a rate of between sixteen and eighteen a minute. When the casualty begins breathing for himself, continue giving assistance at the natural rate until breathing is normal. Then place him in the coma position.

CHEST COMPRESSION

Check that the casualty is lying on a firm surface. Kneeling alongside, locate the bottom of the breastbone. Measure the width of three fingers up from this point and place the heel of one hand on the bone. Lay the other hand over the first. Keeping the elbows rigid, lean forward so that your arms are vertical and your weight bearing down on the casualty's chest. Depress the breastbone by 4-5cm (10-12in). Lean back to release the pressure, so allowing the breastbone to return to its original position. Perform fifteen compressions at the rate of about eighty per minute. (Count: one-back, two-back, three-back and so on, leaning forward on each number.)

In normal conditions, breathing and circulation take place at the same time. The casualty needs both, so assisted breathing and chest compression must be carried out together. If you are alone, the procedures have to be alternated. As soon as you have given the first fifteen compressions, restore the head to the open-airway position and provide two more assisted breaths. When this has been done, resume this cycle – fifteen compres-

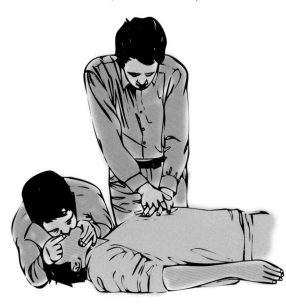

sions and two assisted breaths – for a whole minute. Then check for any heartbeat. If none is discernible, continue the treatment, checking for a heartbeat every three minutes.

If two active survivors are available they should each provide part of the treatment, one assisting breathing, the other providing the compressions. At the start, give four assisted breaths and follow these with five com-

If no heartbeat can be felt, chest compression is called for, above left. The procedure can be carried out more efficiently if two people are available. One person assists the patient's breathing while the other applies chest compression, interlocking the hands as shown above.

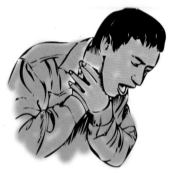

When you or another survivor is choking from an obstruction of the throat, do not insert the fingers. Coughing vigorously is often the best solution. Bend forward from the waist, but do not drop the chin.

pressions. Then establish a pattern of one assisted breath followed by five compressions. Aim at a rate of one compression per second. Each assisted breath should follow the release of the fifth compression without pause. The check for a heartbeat should be made after one minute and then after every succeeding three minutes. Discontinue compression when a pulse is felt. Continue with assisted breathing until the casualty breathes for himself. When breathing and heartbeat are both established, place him in the coma position after checking for other injuries.

CHOKING

Any survivor showing serious signs of choking is in need of immediate assistance. These signs may include being unable to speak or breathe, the skin going pale blue, the casualty clutching at his throat. The condition is usually caused by something lodged in the windpipe which prevents free passage of air to the lungs. Removal of the obstruction is an urgent requirement. A conscious survivor should be encouraged to cough it up. If this is ineffective, check inside the mouth to see if the blockage can be cleared with a finger.

If the choking continues, the combined effect of gravity and slapping should be tried to shake it free. Do this by helping the casualty to bend forward so that his head is below the level of his lungs. Now slap him sharply between the shoulder blades, using the heel of the hand. This may be repeated three more times if necessary. Check inside the mouth and remove the obstruction if it has been freed. If it has not, try to clear it using air pressure generated by abdominal thrusts.

If the casualty is conscious and

THE COMA POSITION

Generally, an unconscious survivor who is breathing has a reasonable heartbeat and is without other injuries demanding immediate attention should be put into the coma position, also known as the recovery position. This position is the safest because it minimizes the risk of impeded breathing. The tilted-back head ensures open air passages. The face-down attitude allows any vomit or other liquid obstruction to drain from the mouth. The spread of the limbs maintains the body in its position. If fractures or other injuries prevent suitable placing of the limbs, use rolled clothing to prop the survivor in the coma position.

upright, stand behind him and put your arms around his waist. Clench one fist and place it with the thumb side against his abdomen. Make sure it is resting between his navel and the lower end of the breastbone. Place your other hand over the fist. Make a firm thrust upwards and into the abdomen. Do this up to four times if required. Pause after each thrust and be ready to remove anything dislodged from the airway. Should the choking still persist, repeat the four back slaps and the four abdominal thrusts alternately until the obstruction is cleared.

An unconscious casualty requiring the thrusts must first be turned on to his back. Kneeling astride him, place the heel of one hand between the navel and the breastbone, and put the other hand on top of the first. Deliver the four thrusts as described above. If the obstruction persists, and the patient stops breathing, it is imperative that you begin to administer assisted breathing and chest compression.

BLEEDING

Bleeding should be stopped as soon as possible. There are three options available.

For persistent choking, alternate abdominal thrusts – pushing the hand into the casualty's stomach – with slaps to the back.

Direct pressure Place a dressing over the wound and apply firm but gentle pressure with the hand. A sterile dressing is desirable, but if none is available any piece of clean cloth can be used. If no dressing is ready for immediate use, cover the wound with your hand. If necessary, hold the edges of the wound together with gentle pressure. Any dressing used should be large enough to overlap the wound and cover the surrounding area. If blood comes through the first dressing, apply a second over the first, and if required, a third on top of that. Maintain even pressure by tying on a firm bandage. However, you should take great care that this is not so tight that it acts like a tourniquet, restricting the flow of blood.

If the wound is large and suitable dressings are to hand, bring the edges of the wound together and use

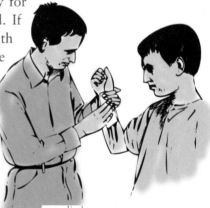

When bleeding is severe, apply direct pressure with a towel, cloth or clothing.

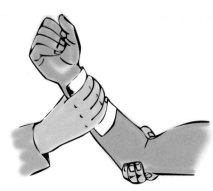

To reduce bleeding, elevate the injured area if possible and apply direct pressure to the wound

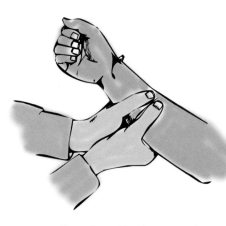

Sometimes direct pressure to staunch the bleeding is not possible because there is a foreign body in the wound. In such cases apply indirect pressure on the relevant vein.

the dressings to keep it closed securely. To arrest the flow of blood from a very large wound, make a pad of the dressing and press it firmly into the wound where the bleeding is heaviest.

The object of this treatment is to slow down or stop the loss of blood until the body's own defences come into play. Blood will clot relatively quickly if the flow is slowed or stopped. However, although a cleanly cut blood vessel may bleed profusely if left untreated, it will tend to shrink, close and retreat into its surrounding tissue. Sometimes these natural methods will succeed in arresting bleeding entirely unaided.

Reassurance and rest also play a vital part in the treatment since they can lower the heart rate and so reduce the flow of blood around the body.

Elevation If there is no danger of any other injury being aggravated, it is best to raise an injured limb as high as is comfortable for the casualty. This reduces the blood flow in the limb and helps the veins to drain the area, and in doing so assists in reducing blood loss through the wound.

Indirect Pressure If a combination of the two procedures described above does not succeed in stopping bleeding, the use of appropriate pressure points must be considered. It is necessary to recognize the type of external bleeding, because pressure points can only be used to control arterial bleeding. Arteries carry the blood outwards from the heart, in pulses of pressure. At this stage, the blood has been oxygenated and filtered of its impurities. Note therefore that arterial bleeding is revealed when bright-red blood spurts out in time with the heartbeat. By contrast, blood which comes from the veins flows out steadily, with less pressure, and is darker red.

A place where an artery runs across a bone near the surface of the skin constitutes a pressure point. There are four pressure points readily available to control heavy arterial bleeding – one in each limb. Those in the arms are on the brachial arteries. These run down the centre of the inside of the upper arm. Pressure points in

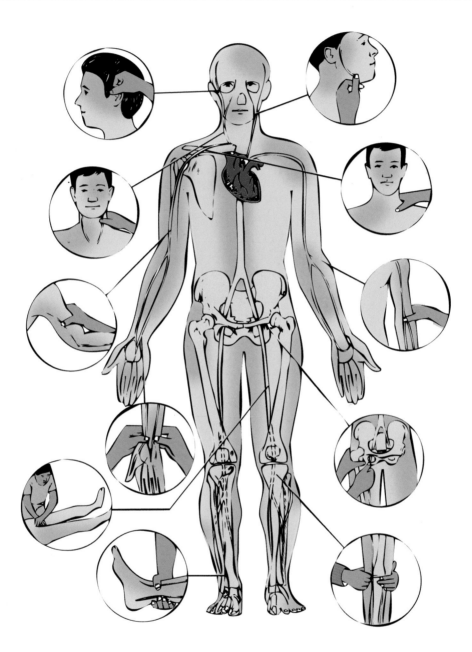

the legs are on the femoral arteries, which run down the inside of the thigh. Other pressure points can be found in the centre of the groin, and can be compressed against the pelvis. This is easier to do if the casualty's knee is bent. When using pressure points to control

There are certain points in the circulation where applied pressure will staunch the flow of blood. Pressure points are hard to find, but experiment until the blood flow reduces.

bleeding, make full use of the opportunity to dress the wound more effectively.

Apply pressure to these points as follows:

1 Locate the fingers or thumb over the pressure point and apply sufficient pressure to flatten the artery and arrest the flow of blood. Then dress the wound again.

2 Maintain the pressure for at least ten minutes, to allow time for blood clotting to begin Do not exceed fifteen minutes or the tissues below the pressure point will begin to be damaged by the deprivation of arterial blood. It is essential to release the controlling pressure after fifteen minutes.

Try to prepare mentally for the possibility that you may yourself be injured, conscious and alone:

1 Lie down and rest – out of the wind if possible.

2 Apply direct pressure to your wound. Put a dressing, improvised or otherwise, on it.

3 Tie on a bandage tight enough to maintain firm pressure without restricting circulation

4 Elevate the injury if possible. Keep as still as possible to relieve pain

FRACTURES

A bone fracture should be suspected if any or all of the following signs are present:

- Difficulty in movement of any part of the body.
- Increased pain when movement is attempted.
- Swelling or bruising accompanied by tenderness in the area of the injury.
- Deformity or shortening of the injured part.
- Grating of bone heard during examination or attempted movement.
- Signs of shock.
- The survivor having heard or felt a bone break.

SPLINTS

The only treatment available in a survival situation is immobilization of the fracture. Unless some other

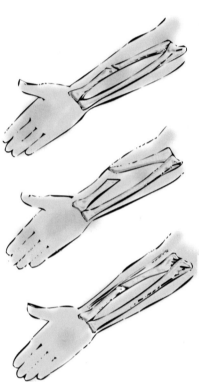

A closed fracture occurs under unbroken skin, and can be complicated by damage to an artery or vein or an organ. An open fracture breaks through the skin, and can have the same results. In either case, when there is injury to a vein, artery or an organ, the fracture is described as complicated.

immediate danger threatens, splints must be applied to the casualty before he is moved. In any case, handle him with the greatest care to avoid further pain or additional injury. If there is a wound associated with the fracture, remove the clothing in the immediate area and treat the wound before fitting splints.

Splints can be improvised from sticks, branches, suitable pieces of wreckage or equipment – even a tight roll of clothing or bedding. Pad the splint and fasten it so that it supports the joints above and below the frac-

Immobilizing a broken arm to speed healing and avoid further strain can be done with a large triangle of cloth. Pad the sling with rag or leaves.

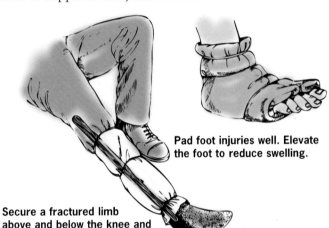

Pad foot injuries well. Elevate the foot to reduce swelling.

Secure a fractured limb above and below the knee and at the ankle.

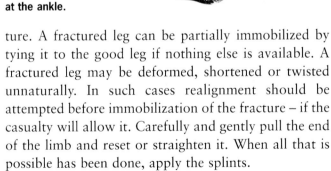

A splint to support a broken arm can be made with a rolled-up newspaper or magazine. In a cold climate, never use metal to splint a limb.

ture. A fractured leg can be partially immobilized by tying it to the good leg if nothing else is available. A fractured leg may be deformed, shortened or twisted unnaturally. In such cases realignment should be attempted before immobilization of the fracture – if the casualty will allow it. Carefully and gently pull the end of the limb and reset or straighten it. When all that is possible has been done, apply the splints.

The only other help possible is to raise the injured part to cut down swelling and discomfort, and to treat any symptoms of shock. The casualty then needs rest.

CONCUSSION AND SKULL FRACTURE

If a survivor is even briefly unconscious, if clear or blood-tinged fluid is coming from the ears or nose, or

An violent blow to the head will often cause dizziness, coma and bleeding from ears, mouth and nose. Lie the victim down and ensure he stays warm. Do not let him stand or try to move around.

if the pupils of his eyes are unequal in size or unresponsive, then skull fracture or concussion should be suspected. If he is unconscious, his breathing and pulse should be monitored. If they are normal, he should be placed in the coma position. If he is conscious, place him in a reclining position with his head and shoulders supported. In either case, keep the casualty warm and handle him gently.

BURNS

The immediate aim when you are treating burns of varying degrees of severity is to lessen the ill-effects of the excessive heat. Do this by gently immersing the injured part in cold water or slowly pouring cold water over it. Persist with either treatment for ten minutes- or longer if the pain is not relieved. Cooling in this way will stop further damage, relieve pain and reduce the possibilities of swelling or shock. It is also important to offer reassurance to the victim.

A burn opens the way for infection to enter the body, and for this reason a dressing should be applied. A sterile non-fluffy dressing is best, but any suitable piece of clean material will do. Dressings and bandages can be made fairly sterile by boiling, or steaming them in a lidded container. Scorching of material will also kill most germs.

A solution of tannic acid will assist in the healing of burns. Tree bark boiled for as long as possible, will provide this. Any tree bark will yield some tannic acid, and while oak is the best source, chestnut or hemlock are good alternatives. As the water boils away, replace it with more, adding extra bark if possible. A strong solution of tea will provide the same assistance. Do not use either solution until it is cold.

If any restricting clothing or other item is being worn near the burned area, remove it before any swelling develops. Do not touch the burn, nor use any form of adhesive dressing. If any blisters form, do not break or drain them. They are a natural protective cover for the

Treating burns should be a combination of covering the afected area quickly with a clean dressing and seeking to lessen the victim's shock with a soothing approach.

injury and should themselves be protected. If burns or scalds are severe, lay the survivor in a comfortable position as soon as possible. If he is unconscious, place him in the coma position at once.

SUNBURN

The type of burn most likely to be encountered in a survival situation is sunburn. Overexposure to direct sunlight, especially when combined with persistent wind, can produce serious burning. Skin, wet with sea water or sweat, is similarly at risk. If sunburn does occur, protect the survivor from further exposure. Treat the area with tannic acid solution (or ointment if available) or with cold water if this is in plentiful supply. Then cover the affected area with a dressing. Keep the dressing in place unless it is essential that it is removed. Provide the survivor with plenty of fluids (or as much as possible) and rest the burned area.

FROSTBITE AND HYPOTHERMIA

FROSTBITE

Exposure to temperatures below freezing, especially if the weather is also windy and/or wet, involves continual risk of hypothermia and/or frostbite. Windy weather increase the risk of both these problems

Remove the frostbite victim's footwear and place his foot in a warm, sheltering part of your body – in the armpit or groin or on the chest.

because the cooling effects of cold air are markedly increased by its movement. Air moving at 48kph (30mph) and having a temperature of -20°C (4°F) has the same chilling effect as air at -40°C (-40°F) moving at only 8kph (5mph). Wet conditions increase danger because wet, cold air is a better conductor of heat than dry, warm air and can therefore carry heat away from the body more effectively.

Take special care of hands and feet. They are the limits of circulation and can lose heat very rapidly. Do everything possible to ensure that the fastenings at wrists, ankles, neck and around the waist are efficient without restricting the circulation of blood. Keep hands under cover whenever possible, warming them under the armpits or between the thighs when necessary. If toes are nipped by frost, warm them against a companion if possible. If you are alone, warm them by wriggling, moving the feet, and by massage.

The risk of frostbite demands particular vigilance, since it can occur without your being aware of it, and

IMMERSION FOOT

When the feet are immersed in cold water for prolonged periods the condition known as immersion, or trench, foot is a frequent result. The feet swell, become white and numb, and the skin may become broken and the flesh ulcerated. It should be kept in mind that this condition does not only develop in extremely cold water. It can happen in water with any temperature below 15°C (59°F) – well above freezing point. If the feet are wet for a lengthy period, check their condition frequently. Immersion foot can be prevented by keeping the feet out of contact with water. Wear sea-boots if these are available. If your socks are wet, remove and empty the boots. Wring out the socks and replace them as quickly as possible. Rub the feet for five to ten minutes periodically, or keep the feet and toes moving. If the condition does arise it should be treated as follows:

1 Dry the feet very gently. Do not rub the skin.

2 Apply an antiseptic cream to any area where the skin is broken.

3 Protect the feet with bandages loosely applied.

4 Keep the body warm, but allow the feet to warm up as slowly as possible and elevate the legs.

5 Do not allow the casualty to walk on damaged feet.

because, besides being a dangerous condition in itself, it can lead to gangrene. There may be a feeling of 'pins and needles' in an affected part. Stiffness and numbness are equally likely to be the only symptoms. Both of these indications will be followed by a greyish or whitish colour to the skin in the affected area.

It is important to check exposed skin areas frequently – in particular the face and nose. If any frostbitten areas are found they should be slowly and naturally warmed. The best method is by skin-to-skin contact – for example, hands in armpits.

Warm water (use the baby-bath test – that is, you should be able to keep your elbow in it) can be used to provide gentle warming. Provide shelter as soon as possible, but in any case insulate the survivor against further loss of body heat, using blankets, extra clothing and other suitable materials. Provide hot food and drink as soon as is practicable.

In cold conditions it is wise to check feet for signs of frostbite at frequent intervals.

If frostbite is detected:
- Do not rub or massage the area affected.
- Do not apply snow or ice. This treatment is dangerous.
- Do not employ hot stones or expose the affected area to a fire.
- Do not give alcohol to drink.
- Do not allow a survivor to walk using a foot that is recently frostbitten.
- Do not break or open any blisters which may appear.

HYPOTHERMIA

When the body loses heat more quickly than that heat can be replaced it is suffering from the condition known as hypothermia. Among the conditions most likely to produce an increased risk of hypothermia are cold, wet weather, wet clothing, immersion in cold water, exhaustion, inadequate clothing and shortage of food or drink. Hypothermia is not an easily diagnosed

Making faces may look silly but in Arctic conditions it will help prevent frostbite. Never rub or massage exposed skin.

condition. It is important, therefore, to keep a special look out if you are subject to any of these conditions.

Signs of hypothermia include:
- Paleness and severe, uncontrollable shivering.
- Being abnormally cold to the touch.
- Muscular weakness and fatigue.
- Drowsiness and dimming of sight.
- Diminishing heart rate and breathing.
- Eventual collapse and unconsciousness (extremely serious).

Cold and wet can kill a sick or injured person very quickly. In such conditions use the nearest available shelter – behind a rock or wall or in a natural hollow. Do not stimulate the body with hot water, fire or friction. The patient will benefit from the warmth of another person in a sleeping bag. Sharing a bag also helps ward off hypothermia in people who are not sick.

In addition to these signs, perhaps the most striking indication of the onset of hypothermia is a marked change in personality of the sufferer. An extrovert may become an introvert. Aggressiveness may change to submission, or vice versa. What is certain is that hypothermia can prove deadly unless it is treated without delay.

The treatment of hypothermia is centred on stopping the loss of body heat and replacing lost warmth. To this end, the following steps must be taken:

1 Provide shelter from the wind and cold as soon as possible.

2 If dry clothing or other covering is available, use it to replace any wet clothing. Replace wet clothing in stages, uncovering as little of the body as possible at any one time. Allow even that part to remain uncovered for as brief a period as practicable.

3 If no dry clothing is available, leave any wet garments on the casualty and cover these with additional insulation against the cold. Add a final, waterproof layer. A metallized emergency blanket is ideal for this purpose, since it is windproof and waterproof as well as reflective of radiated body heat.

4 Provide body warmth – another healthy survivor is a good source.

5 If the casualty is conscious, give hot food and drink.

In the event of hypothermia, bear in mind also the following important points:

- Proceed with treatment even if breathing and heart-beat are undetectable. If this is the case, assisted breathing and chest compression will be necessary. Do not assume death from hypothermia has occurred unless normal body temperature has been achieved and the casualty still does not revive.
- Handle the survivor gently. Frozen skin and flesh are very easily damaged.
- Do not rub or massage the affected area to stimulate circulation.
- Do not warm the casualty too quickly.
- Do not permit the casualty further exertion.
- Do not give the casualty alcohol.

BITES AND STINGS

No survivor should make the mistake of thinking that the greatest danger comes from the biggest animals. The vast majority of big game, snakes and other reptiles want to avoid you at least as much as you want to avoid them. The major – and very grave – threats are presented by some of the smallest forms of wildlife. They can be carriers of debilitating and often fatal diseases, and can convey serious infection when they bite. Every survivor has several lines of defence against disease transmitted in this way and must make the best use of every single one.

INSECT BITES

Any insect bite is potentially dangerous. Mosquitoes, while not particularly dangerous in the Arctic and temperate regions, can be deadly in the tropics. They can carry malaria, yellow fever and filariasis (which leads to elephantiasis). Do everything possible to ensure protection against their bites. Among the most effective precautions are the following:

Remove a lodged bee or wasp sting with tweezers, then run cold water over the wound. For a bee sting, rub on soap or wash the area gently with vinegar or lemon juice.

- Use mosquito netting or repellent constantly if it is

available. If not, cover any exposed skin with hand-kerchiefs, parachute material or anything else to hand. Even large leaves will help.

● Wear full clothing, especially at night. Keep trouser legs tucked into the tops of socks, and shirt sleeves into gloves or other improvised covering for the hands.

● Smear the face and other exposed skin with mud before bedding down for the night.

● Select rest sites or camps which are clear of and higher than swampy ground or stagnant or sluggish water, since this is where mosquitoes breed.

● Establish a slow, smoky fire to windward of the camp-site. Keep it burning to drive insects away.

● Scatter a ring of ash around your bed space, as this will deter most crawling insects.

● There is no immunization against malaria, so any anti-malarial drugs must be used as directed for as long as they last.

Other small creeping, crawling and flying hazards include:

Sandflies These carry and transmit sandfly fever, which has to be treated in the same way as malaria. They are too small for ordinary netting to be effective, but they rarely rise above 3 metres (10 feet) from the ground or fly in moving air.

Insect bites and stings can be a problem for the healthy person, but in a survival situation your resistance to their effects will be much lower than usual. If you are stung by a swarm try to find water, immerse yourself and cover your face and body with mud.

Bee

Hornet

Wasp

Flies There are many types of fly that carry diseases, and the conditions they transmit to humans vary widely from species to species. Protective measures employed against mosquitoes will also usually work against flies.

Bees, Wasps and Hornets All of these are very dangerous if aroused. Nests are generally brownish oval or oblong masses, found on tree trunks or branches some 3-10 metres (10-33 feet) above ground. Avoid them if possible. If a swarm is disturbed, and you are a few metres distant, sit still for five minutes or so, then crawl away slowly and carefully. Should you be attacked, run through the densest undergrowth you can find. This will beat off the insects as it springs back. Immersing yourself in water is another useful defence.

Ticks Carriers of typhus, these oval, flat insects should never be pulled off the skin or squashed while biting, otherwise the heads may remain embedded and become sources of infection. Smoke, iodine, petrol or paraffin applied to their bodies will relax their grip.

Ants Tropical ants can bite severely, and attack in great numbers. They and their nests are best avoided and left undisturbed.

Lice Check your clothing frequently for these pests, which, like ticks, carry typhus. Remove them with delousing powder if it is available. If not, then boiling the clothes or exposing them for a few hours to direct sunlight will remove lice. If you are bitten, wash the affected area with strong soap or a weak antiseptic solution. Do not scratch a louse bite. Doing so will infect the bites with louse faeces, and this can lead to typhus. As a general rule, the less scratching the better, even though it may be very difficult to restrain yourself.

Scorpions and Centipedes These animals are seldom seen, even though they are common, since they shelter under fallen trunks, stones or rocks. They may also seek alternative shelter in discarded bedding, clothing or boots. This is when they are most dangerous. Always shake out bedding, clothing and boots before use. The creatures will not normally attack unless they

Both scorpions and centipedes can inflict a nasty sting. The first of these creatures can kill, but is unlikely to attack unless it is disturbed.

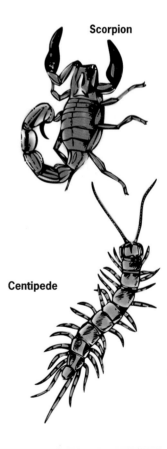

Scorpion

Centipede

A deadly species, the black widow is found, along with other poisonous spiders, In tropical regions.

are disturbed, so use care when moving rocks or stones or handling dead logs. Their stings are painful, but only the larger species are likely to be fatal. Cold compresses will lessen the pain caused by a sting. If you brush or knock the creature away, do it in the direction of its movement.

Spiders Among spiders, only the black widow and its tropical relatives present any life-threatening danger to the survivor. All are dark in colour, with red, white or yellow spots. A bite from any poisonous spider should be treated in the same way as a snake bite.

Flukes and Hookworms These are found in many tropical areas, in sluggish, stagnant or polluted water. Do not drink from such sources (unless the water is boiled), for they can penetrate the skin on contact. Even paddling or bathing in infested water can be dangerous. Once in the bloodstream, they can cause serious diseases, such as bilharzia. They are not found in salt water.

Leeches Lowland forests, tropical and subtropical, are infested with leeches, especially after rain. Their bites cause discomfort and loss of blood, and they also open possible entries for infection. In such conditions, or when wading through swamps or sluggish water, check for these pests every few minutes, since you may not feel their bites. Flick off any which have not yet got a hold, but never pull off a leech, as it jaws will remain in the wound, causing irritation and infection. Salt, ash or a glowing ember or cigarette will make them drop off. If you are smoking, collect unburned tobacco and wrap it in a piece of cloth. When moistened the pack can be squeezed, and it will produce a nicotine solution which is an effective de-leeching agent.

Treat a leech bite by gently squeezing the wound to ensure that it is clean. When the wound is left alone, the bleeding will soon stop. Leave the blood clot on the bite as long as possible. Trousers tucked into tightly laced boots offer good protection against leeches, especially when you are wading. Otherwise leggings should be improvised if at all possible.

Never put your face into water to drink, or a small leech may get into your mouth, nostrils or throat. If this happens, gargle or sniff up very salty water to get rid of them, since they can cause serious infections.

SNAKES

Most survivors are afraid of snakes – or at least the idea of snakes. In fact, our fears are very much exaggerated. Less than 10 per cent of all snakes are dangerous, and almost all of those will do their best to get out of your way if they can. Nevertheless it is essential to avoid alarming, trapping or cornering a snake unintentionally, for when provoked many species will strike with lightning speed. Normally, however, they move very slowly, and can be outran by a man. If you are in an area, temperate to tropical, which has a snake population, treat them with respect by taking the precautions described below.

If you, or any other survivor, suffers a snake bite, your reaction must be rapid, but *without panic*. The widespread, unreasoning fear of snakes contributes in large part to the threat they offer. The two major aims of snake bite treatment are:

1 To reduce the amount of venom entering the body, keeping it below a non-fatal dose if possible.

2 To reduce the speed with which any venom circulates through the system, so that the body has its best chance

In snake country move with great caution – because intrusion on their territory can cause some species to attack.

In the case of snake bite, lie the patient down and keep him very still. Use a restrictive dressing on the heart side of the bite – but do not make it so tight as to restrict circulation. If no medical treatment is available, lower the bitten area in relation to the rest of the body. Give reassurance and rest, but no drinks.

Survival tip
A tourniquet should be used only when all other efforts to stop bleeding have failed. Over-tightening a tourniquet or leaving it on too long risks severe damage or even loss of the limb. Reduce the tourniquet's pressure every few minutes.

to deal with it as it is absorbed.

A moment's consideration will make it clear that any form of fear or panic – especially if violent exertion is involved – will instantly increase the heart rate and therefore the speed of the blood's circulation. It cannot be too strongly stressed that rest and reassurance are high on the list of priority actions in this situation. While this is being given to the casualty, the site of the puncture should be located and copiously washed with water. Do not cut the wound in any way, as this will merely open further channels through which venom can enter the body. And never attempt to suck the venom out of the wound, because the lining of the mouth is able to absorb many substances with ease.

Use a restrictive bandage. Apply it from above the bite, wrapping it downwards towards the puncture site. It should be applied tightly enough to stop the return of venous blood, since this is what will carry the venom into and around the body. But it must not stop the arterial blood supply to the area The correct tightness of the restrictive bandage can be checked by ensuring that (a) there is still a feeble pulse below the bandage, and (b) that the veins below the bandage are distended. The bite will bleed after the bandage has been applied, but this is no cause for alarm. The escaping blood will very probably carry out with it some of the venom from the wound.

The next step is to make sure that the bite is as low as possible compared with the rest of the body. If prac ticable, put a splint on the limb. Immobilizing it will lessen the possibility of any muscle movement exerting a pump-like action on the veins. Then immerse the part in water – the colder the better – as this will further slow down the return of blood.

Reassurance should be constantly given, and the fear of death dispelled as far as possible. It will also lower the risk, and therefore the seriousness, of shock. If fifteen minutes pass and no pain or swelling of the bitten area is felt, nor headache or dryness of the mouth, then the bite is not poisonous.

DANGERS FROM THE SEA

The sea also contains its share of dangerous animals which can threaten the survivor.

Sharks Universally feared, sharks are in fact unlikely to attack unless provoked. They are curious, however, and will investigate any objects in their vicinity. If you find yourself forced to swim through shark-infested water, follow these recommendations to avoid stimulating their curiosity.

● Be as quiet as possible.
● Remove any bright or shiny objects such as jewellery or a watch, as these may look like small fish to a shark.

AIDS

The risk posed by the AIDS virus is particularly great if a survivor is injured and the emergency has occurred in a poor country or in one in which AIDS is widespread. The seriousness of any risk of becoming infected with the virus will depend on the nature of the survival situation in which the individual finds himself. However, in certain countries hospitalization may well increase the risk of infection. The thoughtful survivor will bear this additional danger in mind when planning travel in any country presenting a high risk of AIDS infection.

● Swim smoothly, with as little disturbance as possible. Avoid any splashing. A steady breaststroke is much better than the crawl in this situation.
● The true danger arises when a survivor is losing blood, for then sharks will attack.

Barracuda These are likely to attack without provocation. They are found in tropical and subtropical waters in and around reefs.

Jellyfish The poisonous Portuguese man-of-war is one of the largest of the jellyfish species, but all can inflict stings which are painful. The major threat, however, is that their stings can induce cramp, which is dangerous to even the strongest swimmer. It is sensible to wear clothing

Survival tip
In a survival situation, small open wounds can be sown up with a suture. Clean the wound and pinch the edges together. Put the first stitch in the centre and work outwards. Adhesive butterfly sutures can be used in the same way.

Not every large fin that shows above the water is a shark. But if you are attacked, try kicking and. punching. If you have a knife, stab at the eyes. Be aware that an injured shark will attract others to the area.

Although a jellyfish sting can be very painful, it rarely proves fatal.

Fish such as this stingray, stone fish and scorpion fish all have stings, but if you are stung by one it will be pure accident – the chances of it happening are slim indeed.

while swimming to gain some protection from jellyfish.

Stingrays These flat fish, which can exceed 1 metre (39in) in length, inhabit warm coastal shallows. A poisonous spine on the tail can inflict a painful wound which may prove fatal if the fish is mature. If you cannot avoid wading in such waters, use a stick to sweep the water ahead.

Other fish There are many other species of fish which can deliver severely poisonous stings – mainly from external spines. The stone fish, toad fish and scorpion fish are three examples. They are to be found among coral and in shallows. The weaver is one example of a European stinging species. Look out for and avoid all spiny, odd-shaped or box-like fish, whether for touching or eating. In particular, treat everything you find in tropical waters, along reefs or in lagoons with suspicion and care until you are sure that it is harmless. Any stings received from a spiny aquatic animal should be treated as for snake bite.

WATER

FILTRATION

LOCATING WATER

Nothing is more important to your survival – in everyday life as well as in an emergency – than water. Without it, everything else you have at your disposal in a survival situation – food, equipment, shelter, fire, and so on – is worthless. Your chances of regaining safety will be greatly increased if you properly understand the human body's need for water and the effects on it of prolonged deprivation of its required daily intake.

The body – itself about 90 per cent water – cannot maintain its efficiency without a regular minimum quantity of water. The amount required varies according to climate and the nature of the activity in which you are engaged. Even in a temperate climate the daily adult requirement is 2.5 litres (4.4 pints). If your efficiency is to be maintained – and your chances of survival improved – this need must be met. In addition, everything possible must be done to make certain that the water is pure.

Whenever water is in short supply, the first step is to conserve the water already in the body. To do this:

1 Cover any exposed skin as soon as possible, since this aids water retention, as well as giving protection against sunburn

2 Avoid energetic work during the hottest part of the day, and whenever your tasks make it necessary to move, do it without hurrying.

3 Eat only a little food if water is not available. Talk only when it is necessary, to avoid drying out the mouth and throat.

4 Drink during the cool of the evening or at night – and then in small sips. Don't gulp water down.

5 If seawater or waste water is available, wet your clothes. This cools them and you, and reduces sweating.

6 Don't smoke or drink alcohol.

Bear in mind, though, that all the above actions, however effective they may be, are only short-term responses to the main problem of shortage of water. Long-term survival demands a good supply of drinkable water, and without it your survival prospects are nil unless rescue occurs. Assess your water supplies and then adhere to a planned and disciplined use of what water you do have. Consider other possible sources of water and make plans or take action to obtain it.

It is important to realize that much surface water, especially if stagnant or muddy, will be contaminated with water-borne diseases and will be extremely dangerous to drink unless it is purified. Never underestimate the health risk posed by impure water. The disease-inducing and other harmful organisms it contains constitute one of the greatest enemies of survival.

Although thirst can kill quickly, contaminated water will bring about a slow death. If your only source of water is impure – or even suspect – do not drink any until it has been filtered and purified.

FILTRATION

The first procedure in making water fit to drink is filtration. This will remove creatures of any size as well as particles of mud, leaves and other foreign matter. A short sleeve, sock or piece of cloth filled with sand can be used to filter water effectively. A section of bamboo plugged at the bottom end with grass and filled with sand also makes a good filter.

If they are available, always use purification tablets after carrying out filtration, carefully following the instructions on their use. If none are available, boil the water for five minutes. Try to produce a boil fierce enough to agitate the water thoroughly, as this will ensure equal distribution of heat.

If the climate is hot and sunny, it is a good idea to set up a solar still (*see* page 110). This will purify water by exploiting the same principle that can be used to obtain it from the ground or vegetation.

Passing water through sand, grass or charcoal will filter it. Afterwards add purification tablets or boil it.

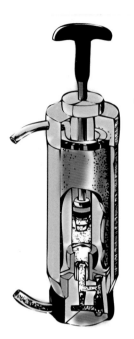

A portable filter will guarantee pure water, but avoid models that are cumbersome.

Charcoal will help to remove unpleasant tastes or smells from purified water if added an hour before it is to be drunk. Don't worry about any small pieces of charcoal that might remain in the water when you come to drink it. A small amount will do you good rather than harm.

PORTABLE WATER FILTERS

Various types of proprietary portable water filter are available. Some are little more than a gravity-fed system which purifies the water as it runs through a series of filters and chemical purifiers. The more expensive models employ a similar system to that in cheaper filters, the only significant difference being that they operate faster in sucking up the dirty water, purifying it and pumping it out.

LOCATING WATER

In populated areas water will be available in all but the most extreme droughts. However, there are many other places where surface water cannot be found, and these are where survival situations are most likely to occur. Below the ground in such areas is an untapped source of water. This, referred to as ground water, is the result of rain which has soaked into the earth, and it is usually pure. The land contours and the rock or soil types will determine how easy or difficult it will be to reach this water. Naturally, its abundance will be determined by the amount and frequency of the local rainfall.

In some situations the indicators of water are

WATER-PURIFYING STRAW

This ingenious device produces safe drinking water directly from the source. Simply place it in the water and suck. The water-purifying straw gives at least one week's use before it needs discarding. It has a remarkable **96 per cent kill** rate for bacteria. Note, however, that it is not designed for use with salt water.

If the immediate area is wet
you might try digging a
swamp filter. The water that is
gathered in this way may be
cloudy, but it will clear if it is
allowed to settle.

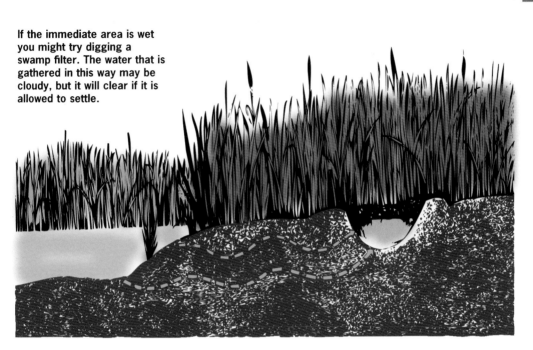

Although the river has dried
up, the bed may yield water.
Dig on the outer bend or
where rock strata meet.

> ### Survival tip
> Desert-dwelling peoples usually know of any locations where water will linger. They will often cover such places in various ways to protect the source from becoming polluted. I was once in the middle of the desert, with nothing but sand for miles all around, when I came upon a battered old suitcase. A closer look at it led to a surprising discovery: it was serving as the cover to a small well. Further investigation revealed that the suitcase held the bucket for drawing up the water!

Dig a well at least 100 metres (109 yards) from the high-tide line. The water will be salty but drinkable.

obvious, while in others they are much more subtle. It is important to be able to recognize the signs.

DESERT

It is in deserts and other arid areas that the survivor is made to realize most dramatically the importance of water. In hot deserts an absolute minimum of 5 litres (8.8 pints) a day is required, and unless this is available the process of dehydration will begin. Whatever water is drunk will slow down the process of dehydration, but as long as body water losses are greater than the intake of water, it will continue.

Any water you have must be used to the greatest advantage. Having taken the steps described earlier to conserve water in your body, bear in mind that the body uses water for digestion and assimilation of food, as well as for the excretion of wastes. So, if you have no water you must not eat. In any case, do not eat or drink for the first twenty-four hours, and then drink only planned amounts at scheduled intervals.

Your overriding need will be to obtain a further supply of water. Never wait until you are without water before you begin searching for, collecting and storing it. When a survival situation arises – whether in the desert or in any other area where water is not readily available – start looking immediately for signs of water. In the desert and arid areas, use these guidelines in your search.

1 Look for any trails – animal or human – that may lead to water. The best supplies come from wells or oases. If you happen to be near either, you may find a trail leading to it.

2 Dry river or stream beds may still have water under the surface. As the surface flow dries up, the water goes underground at the lowest point on the bed. This point will usually be found on the outer edges of bends, and is the best place to dig for subsurface water.

3 Look out for any natural cisterns, which may be found under cliffs, behind projecting rocks or in rock gullies. The presence of animal droppings may be an

indicator of such a place. Any green vegetation shows that there is some water in the ground.

SEASHORE

The nature of the seashore will dictate where it will be best to look for water. On a rocky shore, examine the cliffs for differences in rock strata, for faults – that is, displacement or slippage of rock layers – or green vegetation. Check all these points for water. If the rock layers are not lying level, water is most likely to emerge where they dip.

On a sandy shore, when digging for water at what seems to be a promising place, the problem you are most likely to encounter is contamination of your water by seawater. The best approach is to choose a spot at least 100 metres (109 yards) above the high-tide line. Dig down until signs of moisture appear, and when this is evident continue until the sand is damp. Then wait to see if water seeps into the hole. If it does, do not dig any deeper, otherwise the increased depth may be adding salt water to your supply. Even though it is uncontaminated by seawater, the water obtained in this way will be salty, although it will be drinkable for a short term. Its quality can be improved using a solar still (see page 110).

In an emergency, and provided the fuel is available, salt water can be distilled. Build a small tripod with sticks 1.5 metres (5 feet) in length, then erect this over a small fire. Hang a billycan very close to the fire so that it boils rapidly. Place a large plastic bag over the frame then roll up the bottom 15cm (6in) inwards, clipping it if necessary.

The steam from the boiling water in the pot will condense and run down into the fold. If a plastic bag is not available, cover the pot with a piece of clean cloth. As this becomes saturated with distilled water, replace it with another. To avoid scalding yourself, let the cloth cool for a few minutes before wringing out the moisture into a collection vessel.

COLD AREAS

A simple seashore desalter boils water and collects the condensation in a polythene bag. Ensure that the edge of the bag is turned up so that it traps desalinated water as it runs down.

Compress snow in your hands to obtain water. Eating loose snow will cause dehydration.

SALT

Of considerable importance to the human body, although less so than water, is salt. The average person requires about 10g (0.3oz) each day to maintain a healthy level in the body. Sweat contains salt as well as water, and the salt lost through sweating must be replaced, otherwise heatstroke, heat exhaustion and painful muscular cramps will occur.

The first signs of salt deficiency are a feeling of sudden weakness and a hot, dry sensation all over the body. Resting and drinking a mug of water to which has been added a small pinch of salt will eliminate these symptoms very quickly. In dry desert or sweaty jungle conditions it is advisable to add a small amount of salt to all your fluid intake.

Areas of extreme cold include high mountains and plateaux as well as the polar regions. Finding a dependable water supply may seem a minor problem if you are surrounded by snow and ice, but in reality the task can present considerable difficulties. However, the intense cold does assist the survivor by ensuring that most of the rain or snowfall remains available on the ground instead of running away or soaking down into it. The same cold makes it necessary to produce daily a considerable amount of heat to melt the snow or ice so as to obtain a water supply.

If you must eat snow to quench your thirst, compress it in the hands until water drips from the snowball into your mouth. Do not eat loose snow, as it can cause dehydration. Ice is a better source of water, since it will melt more quickly because it contains little or no air. Crushing the ice into the smallest possible pieces will further reduce the amount of fuel needed to melt it. Blue ice, obtained from a glacier or iceberg, is free of salt and will provide good drinking water.

As soon as any snow or ice has melted, use it as quickly as possible. Make up hot drinks or cook food. Be sure to consume all the water, as there is no point in burning valuable fuel just to allow the water to freeze again. Remember that freezing does not purify water, so that any melted ice or snow that is of suspect origin should be purified before you drink it.

If you are planning to stay where you are for a few days, it is worthwhile setting up an improvised snow melter that exploits the warmth of the sun. If you have one, open up a black plastic bin liner or plastic sheet and make a trough on the ground that slopes down gently towards a water

A lot of heat is required to convert surface snow into water. Use a plastic-bag snow melter over a small fire that is kept burning for a long time.

collection vessel at one end. Support the plastic sheet a few centimetres off the ground with pebbles, rocks or sticks.

When the plastic has warmed, loose snow scattered on it will melt and run down the slope into your container. If no plastic sheet is available, a large flat rock of convenient shape can be used, provided it is placed so as to catch the sun's rays. This will greatly reduce the amount of effort and fuel required to melt snow. Additionally a small fire can be lit under the rock.

Sealing off part of a plant with a plastic bag will produce moisture suitable for drinking.

CONDENSATION

A useful method of obtaining water in a survival situation is through condensation. One way of exploiting this natural principle is to make a vegetation still. All vegetation draws water from the earth and distributes

Drinking water can be obtained by covering a whole plant or shrub with a plastic survival bag. Make a depression in the ground and press the plastic into it to collect the moisture exuded by the plant.

An alternative to collecting transpired water produced by covering part or the whole or a plant is to place foliage in a plastic bag, which should then be sealed tight.

it to the foliage. It is a simple matter of sealing off a section of this foliage to produce moisture. Make sure that you choose a healthy-looking green plant, then cover it with a plastic survival bag, tying the bag's neck around the plant's base. Dig a small depression near the plant and press the plastic down into it to form a collecting point for the moisture which the plant transpires. An alternative, producing a little water, is to gather fresh foliage and seal it in a polythene bag.

Survival tip
Black polythene is the best choice for melting snow, but for distilling water clear plastic is best.

SURVIVAL STILL

If no water can be found in any of the locations discussed above, or if any water found is impure, a drinkable supply can be obtained by the use of a solar still. This is a simple device which will produce water almost anywhere. To make it, a clear plastic sheet of about 2 square metres (2.4 square yards) is required, together with a water container. A plastic drinking tube

Dig a solar still in a place without shade. The sun raises the temperature of both the air and the soil, producing vapourization of ground moisture. This drips into the vessel at the bottom of the still.

Survival tip
If you are injured and unable to move, wrap a piece of cloth around the base of a tree so that a corner of the cloth runs into a water bottle. This set-up will collect moisture released by the tree.

about 1.5 metres (5ft) long is desirable.

Dig or scoop out a hole in the ground, about a metre (3ft) across and some 75cm (30in) deep in the centre, where the collection vessel is placed. Fasten one end of the tube in the container, spread the plastic sheet over the hole, bringing the other end of the tube out under its edge. Secure the edges of the sheet with equally spaced stones. Put a rock or weight on the centre of the sheet so that it sags into an inverted cone. The centre should be allowed to drop about 35cm (14in) below the horizontal. Now put soil or sand all around the edge of the hole to secure the sheet and seal the hole off from the atmosphere.

The sun's rays will pass through the sheet, warm the ground underneath, and evaporate any water present. This will result in the air trapped below the sheet becoming saturated. The water vapour will begin to condense on the underside of the plastic, and the droplets will run down along the sheet and fall into the container. This form of still will usually produce at least half a litre (0.9 pint) of water in twenty-four hours, though very dry desert conditions may yield less than this. In better conditions 1.5 litres (2.6 pints) may be expected.

The following points must be carefully observed during construction of the solar still:

● Make sure that the sheet does not touch the sides of the hole at any point, or the water will be lost back into the ground.
● Make sure that the sheet is clear of the collection vessel, otherwise some water will run down its outside and be lost.
● Check that the seal around the edge of the hole is complete and airtight.

Production of water can be increased if any fleshy plant material is available to be sliced or broken and used to line the hole. Urinating into the hole, avoiding the collection vessel, will also improve production. In this way daily output can be increased to 2 litres (3.5 pints) – even in the desert.

Water can be collected by slicing through a banana plant about 30cm (1ft) from the base. Enjoy the fruit with it!

PLANTS AS A SOURCE OF WATER

Some plants will yield water – or at least drinkable sap – and should be kept in mind for use in an emergency. A few well-known examples are given here, but further detailed study of the plants that are common to the region in which you intend to operate would be worthwhile. Do not drink milky or coloured

plant juices, except for coconut milk and the juice of the American barrel cactus.

Many types of vine are a good source of drinking water and it is always worth investigating them. Cut the vine almost right through as high up as you can comfortably reach. Then sever it close to the ground. Hold this lower end so that any water available will drip into your mouth or collection vessel. When the flow stops, repeat with another vine.

Some plants – bamboo stems are an example – catch or hold water. Try shaking old bamboo stems. If water is heard moving, pierce the stem just above each joint in turn to release it. Trees which will yield water in useful quantities include the bromeliads of the pineapple family, found in tropical America; the umbrella tree of tropical West Africa; and the baobab tree, which occurs in Africa and northern Australia.

Many cup-leaved plants act as water collectors and can be used to quench the thirst. Likewise a cut vine will release water directly into your mouth. Each time it dries up, cut off a piece about 30cm (1ft) long.

7

SURVIVAL SHELTERS

A careful assessment of the survival situation in which you find yourself will make it possible to decide on an order of priorities. One of the key decisions you will have to make is where the need for shelter lies within that order. In most survival situations finding shelter will be a matter of urgency. Even in a temperate part of the world it will usually be a need that you cannot afford to overlook.

The most dangerous weather conditions include cold, wind, rain and snow. It is essential to protect yourself against these, as each of them is a factor which brings about hypothermia – a drop in the body temperature below the normal level. Exposure to one or any combination of these conditions can rapidly produce deadly results long before any shortage of food or water would take effect. Conversely, even in warm summer weather or hot climates, shelter from the sun is needed in order to avoid overheating of the body. Prolonged exposure to high temperatures may not effect the survivor as quickly as loss of body heat, but it can still have fatal consequences by causing a very rapid loss of body fluids.

CHOOSING A SITE

There may be temporary shelter to be found among the natural features immediately surrounding you or nearby. Seek it in or near trees, thick bushes or natural hollows. If on close inspection they seem safe, make use of caves, rock overhangs or any stable form of natural shelter. Never waste precious time and energy constructing a temporary shelter or wind-break if nature already provides it.

The climate and the terrain, along with your personal circumstances – for example, whether you are alone or in a group, the physical condition you are in, and the construction materials and tools you have at your disposal – will to a large extent determine the siting of a shelter and the form it takes. But there are some general points worth keeping in mind when considering the task of building a shelter.

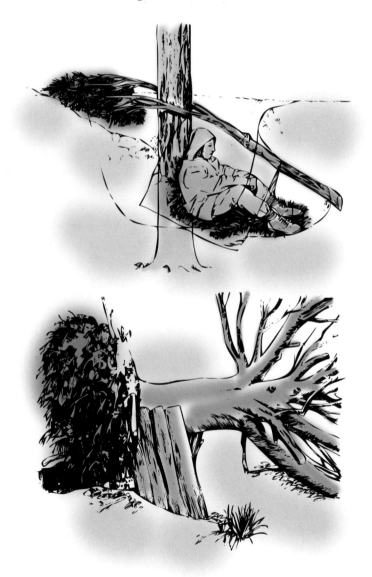

Find cover as quickly as possible. Take into account your state of health and any construction tools available. The most important considerations are insulation from the ground and protection from the weather. Don't attempt to build anything too complicated until you have the time.

A natural shelter, such as this fallen tree, can be easily improved with logs acting as a windbreak. It is most important to settle on your shelter before the cover of darkness or exhaustion sets in.

If possible, choose the site that provides the most natural cover from the wind that you can find. If no site offering such protection is available, angle the shelter so that its entrance or open side is always facing away from the wind. Oddly enough, a hillside is usually warmer than a valley floor, even though it may be windier. Build the shelter as near as possible to a supply of fresh water, to sources of building materials and, very important, firewood. Any spot that is in a forest and near a fast-flowing stream can provide a site for a desirable, if temporary, residence.

In lowlands be aware of the danger of floods. On the coast keep the tides in mind. In mountainous areas make sure that your prospective site does not lie in the path of possible avalanches or rockfalls. If you are in a forest, look around for fallen trees, which may indicate that it is an area of shallow soil. If the wind can blow one shallowly rooted tree over, it could do the same to others near you. For the same reason, isolated single trees are best avoided. On the other hand, the branches of an isolated tree which has already fallen may well provide a ready-made framework for a sound shelter.

CONSTRUCTION MATERIALS

In survival situations where a ready-made natural shelter is not available, the shelter is usually constructed from a combination of materials possessed by

Tuck yourself into a fallen fir tree. It may be damp, but you can increase your comfort by building a fire.

the survivor and natural materials found at or near the site. Probably the most useful example of the former is any type of sheeting. A groundsheet, plastic sheeting and sacks, canvas and blankets can all be used to provide a windproof shelter. Among the most commonly used natural materials are:

Turf Turf can be used for constructing a shelter in very flat, open areas where trees and shrubs are scarce. Indeed in many countries it is still used as an effective roofing material.

Foliage With foliage it is possible to construct an excellent, long-lasting waterproof shelter. If it is available, use large-leafed foliage.

Stones Where the ground is too hard to dig a shelter, or foliage is in very short supply, it is usually possible, though more time-consuming, to build a good shelter with large, flattish stones.

TERRAIN AND SHELTERS

Different types of terrain provide different features and materials for use in creating a shelter. The most commonly encountered types of terrain and the materials they furnish are examined below.

FOREST

In any forested area there may be large fallen logs to be used. A trough dug between such logs, covered with a roof of branches and foliage, provides a shelter which requires little effort to construct. A single log can be supplemented with a low earth wall or used as the basis of a small lean-to.

The lean-to frame is the most commonly used shelter pattern – probably because it is among the simplest. When setting up the frame, make sure that the roof slopes down into the prevailing wind. The covering can be provided by a wide range of materials, from foliage to plastic sheeting, from a groundsheet to panels from a wrecked vehicle or aircraft. Even blocks of turf can be used. A firm mud or turf layer, when placed on top

Groundsheet

Sacks

Turf

Foliage

Stones

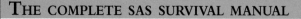

One of the best shelters to build is a simple lean-to. Note the wind direction and then site your frame accordingly, using available branches to construct it.. Cover the frame with foliage thick enough to stop the wind and rain, and then cover the sides to give extra protection.

LAYERED TREE BIVOUAC

A small bivouac-like shelter can be made quite quickly from any small tree. Cut partially through the trunk at about shoulder height until you are able to push the upper part over so that its top rests on the ground. Do not detach this upper portion from the lower. Cut away the branches on its underside and break the upstanding branches on the outside so that they hang down. Thatch the shelter using the foliage cut from the underside.

of foliage, will harden, prevent the shelter cover from being blown away and make it more windproof and waterproof. The sides of the shelter can either be filled in by using a similar combination of foliage and mud or turf, or they can be built up with blocks of turf.

If the survivor has a supply of usable cord or string, a variation of the lean-to can be built that is especially suitable in any area covered with short, bushy vegetation. Cut four or five stakes of the greatest length available. Force their ends into the ground, then bend the tops over and tie them down at an angle of 45 degrees. It may be possible to find a spot where two or three can be incorporated into the framework of the shelter without being cut – that is, they can left rooted in the ground. This will ensure that the shelter has much greater stability.

Twigs and branches can be tightly interwoven to provide a firm, hard frame. More foliage is then added until a complete covering is formed. A layer of light turf or mud will complete the roof.

PARACHUTE TENT

If you are lucky enough to have a parachute it is a simple matter to fashion a bell-tent from it. Parachute material is wind-resistant, and shower-proof as long as it is not touched, but it does not offer protection against heavy rain. First remove all the valuable paracord rigging lines. Tie a good length of cord to the centre of the parachute canopy. Tie a suitable log or stone to the other end and lob the cord over a convenient tree branch. Pull the parachute to its full height and secure the cord to the trunk. It is then a simple matter of pulling out the parachute skirt and pegging it in a circle. The tent can be improved by cutting one side of a panel to form a door – on the lee side. A stove suitable for use inside this kind of shelter is the Yukon stove (see page 148).

PLAINS AND GRASSLANDS

If you find yourself in an area which is covered with grass but where trees are either scarce or completely absent, it is possible to cut turf bricks and build an effective shelter. Making the roof will be easiest if you can find any small sticks or boughs to support a turf roof. An alternative form of roof is any kind of sheeting which can be anchored by the top row of turf bricks. If nothing but turf is at hand, make the shelter small and narrow enough for longer turf strips to be used in pairs, supporting each other, for the walls.

If the grass is long enough, bundles might be suitable to make a thatched roof for the shelter, but in this case try to pitch the roof as near to 45 degrees as practicable to provide a run-off for any water.

If the ground is suitable – that is, soft but not wet – it may be possible to combine the effort of digging or cutting the turf with building a low wall along the edge of the slit trench produced in the process.

The effective height of the wind-

Before bricks, most houses were constructed with a timber frame supporting wattle and daub – a lattice-work of sticks caked with mud. The same method can be used to build shelters on plains and grasslands.

In areas of short bush, a willow shelter can be made by overlapping saplings to form a frame. This can be covered with a shelter sheet, plastic sacking or a parachute.

Turf blocks make a wet but windproof shelter. Insulate the ground as well as possible, since it will inevitably fill with a little water (below).

break is then increased for about the same amount of effort. It is essential to make sure, however, that any rain will drain away from the trench and not into it.

Making a shelter below ground level can also be very helpful in hot conditions, although when the weather is hot the hardness of the ground frequently makes this difficult.

All these shelters, and especially the lean-to varieties, can be improved by the addition of a fire and fire reflector. The fire is best set on a base of green logs, while the reflector is made of interwoven green sticks. Large stones stacked around the back of the fire will also reflect heat. The hot stones can be taken into the shelter at night. With care these can be placed beneath your bed space, where they will continue to emit warmth during the night.

SNOW

There are several types of shelter that are used specifically in snow-covered terrain. However, some of them suffer from significant drawbacks. One type may demand too much time and energy in its construction, especially for a solitary survivor. Another might require a greater depth of snow than is available. Alternatively, very cold, hard-packed snow may not be available for the cutting of snow blocks.

SNOW TRENCH

Even a hole in the snow provides temporary shelter as an emergency measure, and it can be improved to make a simple shelter for one man. If the snow is soft, branches or sheeting will be needed for the roof.

Survival tip
Constructing a snow hole requires a great deal of effort, but you can make it easier by removing the undergarments from the top half of your body and tunnelling into the snow with waterproofs covering both the top and bottom half. Put the dry undergarments back on when the hole is complete.

At the very least a snow trench will protect you from the wind. But with small refinements, such as pine branches for insulation, and a candle, it provides a good night's sleep.

FIR-TREE SHELTER

If in a wooded snow-covered area, by far the most convenient and simplest shelter is to be found under a large fir tree. There will in many cases be a natural hollow in the snow around the trunk of the tree and this will give you a good starting-point for building the shelter. First

A simple snow cave, as illustrated on the opposite page, can be easily excavated and offers snug and secure overnight accommodation.
(1) The entrance porch is dug out first, by removing snow in a vertical trough from a deep and well-drifted snow bank.
2) Continue digging out a T-shaped hole to a depth of about a metre (3ft). The upper part of this shape will form the sleeping area.
(3) Carry on excavating the horizontal sleeping area a further metre (3ft) beyond the end of the vertical entrance. Any snow removed can be used to build up walls around the porch that will give added protection from the wind.
(4) Cut snow blocks large enough to fill the front of the sleeping area.
(5) Leave gaps between the blocks for ventilation.
(6) Crawl in and upwards onto the sleeping area. Dig upwards to create a domed roof and leave the entrance well open for further ventilation. If the wind is too strong the well can be closed with a rucksack.

dig away the snow from the base of the tree and use it to build up and improve protection from the weather either side of the shelter. Cut the low branches on the side away from the shelter to use as bedding or to inter-

weave with the branches on your side to improve the overhead cover. You can build a fire under the tree, but make sure it is at least part of the way around the trunk from your shelter, to stop it melting snow overhead.

Snow cave

This type of shelter requires a depth of snow of 2 metres (6.5 feet) or more, and so it is appropriate in

In snow-covered woods, large fir trees make an excellent natural hollow in the snow. This can be insulated from the ground by cutting away some of the lower branches, at the same time creating soft bedding. Always make your fire on the other side of the tree.

A raised platform inside a snow shelter stops the coldest air – which sinks – from getting in. A second channel is needed for ventilation.

Igloo

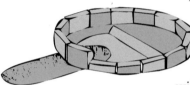

very cold regions where the snow level builds up for long periods. The simplest approach is to dig into a snow drift or cornice. To improve the snow cave, aim to incorporate as many of the features shown in the cross-section above as you can. Make sure that the inside roof is always dome-shaped, or you will wake up in the morning with it on your head.

Note that sometimes this type of shelter is considerably more difficult to make than it might seem, because of the hardness of the packed snow. In fact, without tools other than your hands and feet it may prove impossible.

SNOW IGLOO

If the plan is to remain in one location for more than a day or two an igloo built of snow blocks will provide a good refuge for two or more survivors. It requires tools for its construction – an axe, a knife, and a saw or a spade. It also takes time and effort, as well as care and thought in placing the blocks. However, the combined efforts of two or three people will reduce the difficulty considerably. The blocks must be cut from cold, heavily compacted snow – no other type of snow is suitable. Build up from the base, gradually working towards a point above the middle of the igloo, so that the blocks eventually join overhead.

The secret of building a good igloo is to angle the lowest layer of blocks so that walls slope inward.

Where the snow is too shallow to build an igloo, a snow hive can be made.
1 Pack snow on to a dome of branches covered with a sheet.

2 Once the layer of snow is about 1ft (30cm) thick, carefully remove the branches and the sheet from the core.

1

2

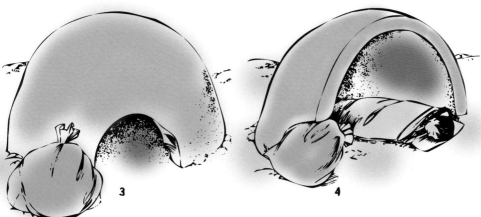

3

4

An alternative method of constructing an igloo is to stamp down an area in the snow and build up a mound of packed snow, then simply tunnel into the mound. This method of construction has the advantage of being quick to construct, and is easier for the novice or the lone survivor.

Take care to lay out the entrance tunnel on the lee side of the shelter. In any snow shelter, build your bed on a raised platform, so as to avoid the coldest air, which will gather at the lowest level.

3 Use your rucksack or something similarly bulky to block up the entrance once you are inside the hive. Remember to leave a small gap in order to allow ventilation.
4 This provides a surprisingly strong night shelter. On two occasions I have built hives and they can stand for seven days, so it is worth taking time to construct a large one.

1 For survival shelter in long-term desert conditions, but where a building material is available, construct yourself a makeshift shelter by digging into a sand-dune.

2 Sand has a habit of drifting and finding its way into every crevice. This can be minimized by using any wood and vehicle and aircraft parts you can find to form the basis of your shelter.

3 The shelter should be covered, if at all possible, to protect you from the wind, sun and night-time cold – a real possibility under the cloudless skies of a desert.

As soon as you have constructed your igloo, place a lit candle in the centre. You will be surprised by the warmth generated by so little a fllame.

DESERT AND ROCKY AREAS

The task of obtaining any form of shelter in desert areas presents several difficulties. However, the possibilities should be given some thought – not least because roughly one-fifth of the earth's surface falls into this category and survival situations frequently arise in such terrain.

The first difficulty is that deserts are places of extreme conditions – extreme heat during the day and biting cold at night. They also vary greatly in their composition, consisting of rock, sand or salt, or any combination of these. Some deserts are plains, others mountainous, still others depressions. Some are totally barren, others have scanty vegetation, while some have a variety of plants. All these variations can occur in combinations which make desert shelter difficult to achieve if you are entirely without material resource.

THE SANGAR

A fortified combat position named after the Persian word for stone, a sangar is one of the earliest forms of man-made shelter. In the context of survival it is simply a windbreak built of any materials available – stones, branches, snow, parts from wrecked vehicle or aircraft, or indeed anything else that is suitable. In the absence of any better alternative, the sangar at least has the benefit of reducing exposure to the chilling effects of the wind. Use a survival blanket, poncho or plastic sheet as a roof to give shade by day and as a blanket at night, unless it is needed for protection against rain or snow.

Shelter from the sun and heat is the main aim when in the desert. Use a groundsheet (or any alternative) to cover a depression scooped out of the ground. This is

A sangar is nothing more than a circle of stones. In the Middle East it provides protection against wild animals as well as the sun. An inner framework, supporting a sheet roof, much improves it.

If two forms of sheeting are available it is best to make a layered roof, leaving a 2-3in (5-7cm) gap. This allows air to circulate and drastically reduces the temperature.

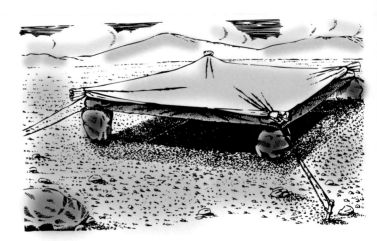

Survival tip
In jungle conditions there is usually an abundance of troublesome and unpleasant insects — on the ground. This means that a good basis for any shelter would be a raised platform. Even if your resources do not make it possible to build a platform big enough to support the entire shelter, it is very important to avoid sleeping directly on the jungle floor.

known as a 'scrape'. If there are any rocks or vegetation, drape the sheet over the rocks or plants. If you have no material help, look for shade or shelter from natural desert features – rocks, rock cairns, caves or ledges. Dry stream beds may offer shelter. These wadi banks, or the sides of ravines or valleys, are worth looking over for crevices or caves.

LONG-TERM DESERT SHELTER
Dig into the lee side of a dune, and make a roof from any material available. (A life-raft from a crashed aircraft is ideal. Use the paddles to support the inflated raft.) Cover the whole structure with any cloth or plastic material to prevent sand infiltrating it. If it is

In mountainous desert areas, there is usually a good supply of caves. Always ensure that your selected cave is unoccupied Form your fire at the deepest point and block the entrance against the wind and wild animals. Conditions can be made more tolerable by using soft sand from a valley floor to form a comfortable bed on rocky ground.

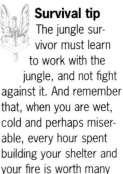

necessary, camouflage the whole shelter with sand, as this will not only aid concealment but will help to keep you cool during the day and warm during the night.

Whichever type of desert shelter you set out to build, remember that your aim is to protect yourself against those aspects of the environment which can threaten your safety. In any area with a hot climate, insects are likely to be a problem. You can gain some protection against the winged species if you are able to erect your shelter on a site which receives some breeze – for example, on a hillside or ridge, or in a location which receives an onshore wind.

Survival tip
The jungle survivor must learn to work with the jungle, and not fight against it. And remember that, when you are wet, cold and perhaps miserable, every hour spent building your shelter and your fire is worth many hours' sleep.

JUNGLE

In the jungle, the survivor should never be short of materials for building a shelter. All such items are likely to be close at hand, but you would do well to select the site for your shelter with care. These are the main factors that will influence your choice:

- The presence of nearby food and water.
- Stable ground away from swamp or infected areas.
- Protection from danger, such as rotting or falling trees and wild animals.

BAMBOO

Bamboo (atap) is one of the most commonly used building materials in survival in the jungle, but gathering it can be hazardous. Care should be taken when cutting bamboo as it grows very densely and in some growths sections are under strain. It is not uncommon for a cut section to suddenly shoot forward and hit you with some force. Bear in mind as well that bamboo is very sharp.

Despite these drawbacks, bamboo is a wonderful material and with the aid of a good jungle knife, you can construct many useful survival items, including the shelter itself, a pole bed, cooking and drinking utensils, and even a serviceable raft. Vines are normally plentiful

Foliage for shelters is easily found in any jungle. Normally just a small portion of one plant will be ample to cover your frame; it also makes excellent bedding.

An A-frame pole bed is usually supported by two larger trees. The tension in the bed – which can be made from plastic feed sacks – holds the bedding poles in place.

and only require pulling down from their branches, but again care should be exercised. Always look up first.

POLE BED

Build your bed first and then construct your shelter over it. A pole bed of bamboo or any small branches covered with palm leaves or other foliage is a real necessity. If any sizeable suitable material is available then you should use that for the base.

HAMMOCK

You can make a hammock if you have a parachute, since this will provide almost the ideal materials. However, do not attempt to make one out of vines since they normally break.

A perfect combination for a shelter in a warm climate is either a bed or hammock slung beneath a poncho or shelter sheet. Providing it does not become too cold, it is possible to live comfortably for many months underneath such an arrangement.

JUNGLE SHELTER

If you intend to stay put for a while in the jungle, it is fairly easy to construct a very comfortable dwelling in a short period of time. However, as always, it is best to

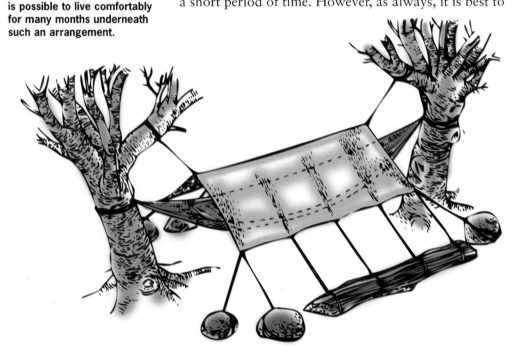

There is an old saying that the jungle is neutral. It will quickly provide you with good shelter, food and water. If you intend to stay put for some time, make the effort to construct as good a camp as you can.

Almost the biggest menace when you are in the jungle is not large animals, but insects. It is essential to sleep on a raised platform. Burning a termite nest keeps flies at bay – but the smoke can be almost as irritating. A layer of mud stops biting insects and can be washed off in the morning.

Use broad leaves to cover a jungle shelter. Large branches can be split and woven into a frame of saplings. Experiment with whatever cover is available in the vicinity.

plan your shelter carefully. The major point to consider is where to position your fire. If the shelter is big enough, and there is no possibility of it catching fire, then inside is best. During the day the fire can be used for its normal functions, notably cooking, while at dusk and during the night it can be used to fend off insects. A good idea is to burn a termites' nest if you can find one, since this will produce a great quantity of smoke which will keep the flying bugs away. However, bear in mind that it is not always clear which is the most evil of the two: the smoke or the mosquitoes. Spreading old ash around your bedding area and the shelter site will also help reduce the amount of crawling companions that gather during the night.

8

FIRE

Fire is one of the essential aids to survival, and therefore the ability to light a fire in difficult circumstances is among the most valuable of all survival skills. The discovery of fire was one of humanity's most important advances, since, in combination with the use of shelter, it allowed early humans to survive in otherwise unsuitable climatic conditions. It encouraged mobility and gave them a choice concerning the environment in which they lived.

It is perhaps because fire has played such a vital part in our history that it continues to play an important psychological role in survival situations. Fire is a source of comfort, and lighting a fire is proof that a survivor can counter at least some of the dangers and difficulties which face him. It also provides a sense of achievement in that he has, despite his emergency situation, replicated some of the most important elements of normal everyday life – the ability to cook food, to feel warm in inhospitable conditions and to see in the dark. In addition, fire can be used to purify water and to sterilize medical equipment, to dry clothing and to generate signals seeking help.

THE ESSENTIALS OF FIRE

Any fire requires three elements: heat, fuel and oxygen. If any one of these is missing, it will not burn. When you are choosing material with which to start a fire, it is helpful to understand that fire is a form of chain reaction. Part of the heat generated by the combustion of any fuel is required to ignite the succeeding supply. The initial supply of heat available to start the fire is usually small – for example, a match burning for only a few seconds. It follows that the starting fuel, in order to be set alight by such a brief flame, must be a material that ignites easily. Such materials are loosely termed 'tinder'.

TINDER

It is essential that tinder for fire-lighting is dry, and it will ignite more readily if it is reduced to fibres, threads or shreds and piled loosely so as to improve ventilation and thus combustibility. Once alight, it will burn quickly. It is therefore essential that, before setting light to tinder, you make certain that a supply of kindling wood is to hand to consolidate the fire before larger pieces of wood or other combustible material are added.

Among the many possible sources of tinder, the following are the most likely to be available in a survival setting:

● Decayed or powdered dry wood
● Catface (the resinous scab found on damaged evergreens)
● Pulverized outer bark (especially of cypress, cedar and birch)
● Coconut palm frond (the fabric-like material at the tree's base)
● Crushed cones from evergreen trees
● Arctic cotton grass or sedge (Eriophorium) (August-September)
● Down from bulrush
● Termites' nest material
● Ferns, moss, grass, evergreen needles and dry fungi
● Any fine dried vegetable fibres
● Scorched or charred cloth (especially linen)
● Pulverized cloth (lint) (scrape to a fine fuzz and use in a loose pile)
● Petrol or jet fuel (caution required)
● Insect repellent
● Oil

(These last three are all best used in conjunction

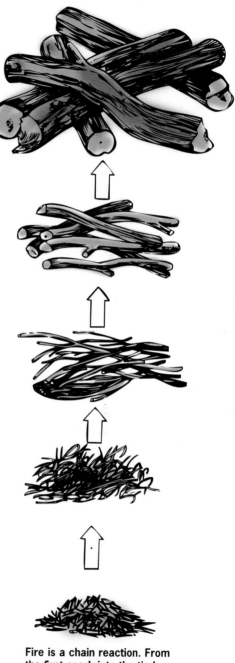

Fire is a chain reaction. From the first spark into the tinder, build your fire slowly with kindling, and then add progressively larger twigs and sticks.

The best kindling can be made by repeatedly scraping along a dry stick with a knife. The shavings this produces readily convert glowing tinder – such as the cotton wool from a field dressing – into the beginnings of a fire.

Rather than making a new fire every time it was needed, our ancestors quickly found it was better to preserve embers from the previous fire. A modern-day version can be constructed in the same way using a ventilated tin can.

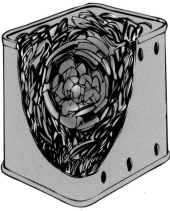

with some solid form of tinder, or poured over sand or an absorbent material.)

● Matted body hair
● Birds' nests
● Rat or mice nests
● Gunpowder (obtainable, with care, from ammunition)
● Bat droppings (dry and powdery)
● Some photographic film
● Charred rope, lint from twine, canvas, bandages or similar fabrics.

KINDLING

The term 'kindling' refers to small, dry twigs and sticks, the first of these being added to a fire before the second. The need for dryness at this stage cannot be overemphasized – the wood you gather for use a kindling should be dead, and as dry as possible. Once burning well, kindling enables a small but hot fire to be established. When this has been achieved you should slowly add progressively larger sticks until you have a fire which will ignite small logs. Even unseasoned logs can be added to such a fire, since the heat available will boil away the sap before the logs burn.

MAIN FUEL

Once your fire is established you can add the main fuel, which is most likely to be wood, as necessary. It will now be possible to burn large pieces of wood such as tree trunks. However, do not waste fuel, since it takes valuable energy to collect it. Collect dry standing wood if possible, since wood found on the ground is likely to be damp. Bear in mind that the harder the wood the longer it will burn. Hard wood also gives off more heat than soft wood and provides excellent hot embers to carry to your next camp-site. Soft woods will burn quickly and give off flying sparks, great for a signal fire but not recommended if you are on the run.

Before you attempt to light your fire, it is essential to

collect sufficient fuel to build and consolidate it. You should then grade it according to its use and stack it in three piles: tinder, kindling and main fuel. It is very important not to make the common error of piling kindling and other wood on to the fire too soon. Doing this is likely to limit the supply of oxygen, so that the fire dies. If you ensure that the fire is well ventilated, it will burn efficiently with small pieces of wood, and these will produce less smoke than large pieces.

Don't use kindling that has been lying on the ground, since it is nearly always damp right through. In wet conditions reach a little way up the trunks of trees to find dead, dry twigs.

To carry burning embers, pack them with small, unburned sticks, dried by the fire. Wrap them in moss, and surround them with bark to make a portable package.

HEAT

The heat required to ignite tinder can be generated in various ways. The easiest method is to use an open flame from a match or cigarette lighter. Sparks from flint and steel, or from an electrical source, can also be used. In sunny conditions a magnifying glass or a parabolic reflector can be used to concentrate the sun's heat on the tinder until it ignites.

MATCHES

A supply of matches is carried as a matter of course by those likely to find themselves in a survival situation, for, along with the cigarette lighter, they are the easiest means of generating flame. They are, however, vulnerable to the effects of damp. Ordinary household matches can be protected from water by dipping each match into molten wax so as to cover the head and half the stick. An alternative method of protection from damp is to spray both the matches and the box with hair lacquer.

PORTABLE FIRE

In cold climates the ability to produce fire to order is essential to survival. If you are on the move in such an environment and there is a limited quantity of fire-lighting materials, it is a good idea to carry your fire with you. Rather than try to light a new fire every time he needed one, early man, it is believed, eventually hit on the idea of making fire portable. He did this by using sticks to transfer glowing embers from the fire into a fireproof container made of bone. Today the survivor can do the same by using an old medium-sized can fitted with wire for a handle. Hard-wood embers are the best source of fire to carry, and should be semi-starved of oxygen by surrounding them with tightly packed charcoal.

Survival tip
Whatever type of matches are used in a survival situation, the aim should be the same: a fire from every match.

LIGHTER

A cigarette lighter can be a life-saver in a survival situation, but it must be used sparingly. Once a fire has been established, one way to economize on a lighter's use is to use the fire to dry out kindling for the next fire. This is particularly advisable if you are on the move. With tinder prepared in this way, the lighter is needed for the minimum time and its fuel is conserved. Bear in mind that if the lighter's fuel does eventually run out, sparks from the flint can be used, in conjunction with tinder, to light a fire.

CANDLE

A candle, however small, will help you conserve matches and prolong the active life of your lighter. With a single match or brief use of the lighter, a candle can be lit immediately. You then have to hand a long-lasting naked flame with which to ignite the tinder. This will prove particularly useful if repeated applications of the flame are needed because you are forced to use slightly damp tinder.

To put this fire-lighting technique to use, scoop out a hole in the ground to the depth of the candle or build a shall shield of stones to the height of the candle. Lay

A sheltered candle burns for a long time, making it an ideal way to ignite damp kindling.

the tinder over the hole or stone shield, and slide the lighted candle underneath. As soon as the tinder is alight and burning well, remove the candle, extinguish

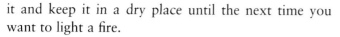

it and keep it in a dry place until the next time you want to light a fire.

Any hot wax should not be wasted. Make use of it to waterproof something, or keep it to improve the combustibility of the tinder for your next fire. Wax itself can be conserved if the candle is set in a small metal container – for example, a metal 35mm film container – so that it works like a night light – which, of course, may be one of its uses.

BURNING GLASS

The use of a burning glass is restricted to reasonably bright conditions, but in such conditions it is an effective aid to fire-lighting.

It may be a purpose-made piece of equipment, such as a magnifying glass, or can be improvised from a lens from a pair of binoculars, a camera, spectacles or a compass. An efficient size for a burning glass is about 5cm (12.5in) in diameter.

A burning glass requires very dry tinder, such as cotton wool. Hold the glass so that the sunlight is focused on to the smallest area possible.

FLINT AND STRIKER

A manufactured flint with an attached striker made of serrated steel is a very useful fire-lighting tool and is available commercially. Struck rapidly downwards, the flint will cause sparks to fall on to the tinder and ignite it. Natural flint is not easy to find, but it is possible to buy it from gunsmiths who deal in weapons that use black gunpowder.

MAGNESIUM FIRE STARTER

One of the better fire-lighting implements on the market, the magnesium fire-starter is highly recommended for inclusion in any survival kit. It consists of an aluminium block impregnated with magnesium and with a bar of flint along one edge. Any sharp edge will produce shavings from the block, and these can be ignited with ease by sparks struck from the flint bar. The magnesium content burns at a heat in excess of 2760°C (5000°F) and will set light to tinder even when it is damp.

Good for at least three thousand fires, a modern tinder-box contains an artificial flint, striker and a magnesium block, tinder shavings from which burn white-hot.

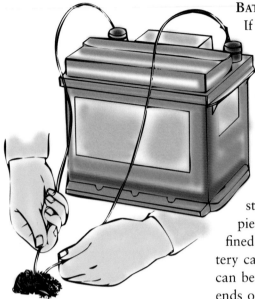

A battery from a damaged vehicle can start a fire. Join wires to each terminal and, holding the insulation, short the battery out across wire wool mixed with a few magnesium shavings.

Survival tip

Most soft aluminium castings in ships and aircraft can be scraped with a sharp tool to produce very fine shavings and dust. In this form they will readily ignite, to produce a very hot flame. Note, however, that the burning time is brief, so you must be ready to take advantage of this fire-

BATTERY

If a battery of substantial capacity is available – for example, from a wrecked or broken-down vehicle – a useful electrical means of lighting a fire is to hand. A wire running directly between the positive and the negative terminals will cause a short circuit and in doing so will become hot very quickly. The thinner the wire, the more quickly it will heat up, so care is necessary when handling it. Even better, if a short length of thin wire – of, say, one or two strands – can be connected to each terminal by a piece of thicker wire, the heating is likely to be confined to the thin section. A wire thus heated by a battery can be used to ignite tinder. Alternatively, sparks can be produced over a pile of tinder by brushing the ends of the heated wire together. In either case it may be necessary to repeat the operation several times before ignition is achieved.

SPONTANEOUS COMBUSTION

Making use of the phenomenon of spontaneous combustion may seem an unlikely method of starting a fire, but it may be possible even in a survival situation, particularly if a vehicle with a water-cooled engine is present. There are two essential ingredients. One is anti-freeze solution, which a vehicle's radiator is likely to contain. The other is potassium permanganate, which is often included in a survival pack for use as a mild antiseptic. When anti-freeze, even if diluted to a slight degree, is added to potassium permanganate, spontaneous combustion can be caused by the heat generated during rapid oxidization.

Place a teaspoonful of potassium permanganate on a sheet of paper or other inflammable material and add two drops of anti-freeze. Then roll the sheet up tightly and place it on the ground and spread tinder over it. Combustion will occur within a couple of minutes. However, too much liquid will slow the rate of heating, and the paper will only smoulder, and will need to be fanned or blown in order to burst into flames. (The tight rolling of the sheet is necessary to ensure that the heat is not allowed to dissipate and so fail to raise the temperature to the necessary 233°C (451°F) – the flash-point of paper.)

INFLAMMABLE MATERIALS

There are many highly inflammable materials which will serve as very effective tinder. Always check your resources very carefully to see if any such materials are available. Gunpowder, which is a mixture of potassium nitrate, sulphur and charcoal in equal parts, is available from small arms or shotgun cartridges. Take great care when trying to extract it.

Another mixture, consisting of equal parts of sugar and sodium chlorate, will give a high heat output, and will be of great assistance if you are trying to light a fire with damp materials.

Warning: It is vital to remember that many man-made materials produce poisonous gases when they are burned. Never burn them in any constricted space – such as a cave – and avoid breathing in any smoke if they are being used as tinder.

FRICTION

While friction is a good source of heat, it is not as suitable in a survival situation as the fire-lighting methods described above since it requires considerable expenditure of energy and time. Both of these precious commodities are best used sparingly when the survivor faces a threat from the elements or a human enemy, or both. However, he must be able to create a fire, if at all possible, with what he has. Even if he has some means of making a fire – matches or a cigarette lighter, for example – these are finite resources. Therefore the time may well come when some other, non-exhaustible method has to be found.

Again it is worth considering the solution hit upon by

Add a few drops of anti-freeze to a small pile of potassium permanganate on a sheet of paper. Roll this into a tight ball and place under kindling.

Only if absolutely necessary, a bullet casing can be cut and packed to light a fire.

primitive man. He created fire from friction – and in some parts of the world people still do. Anyone with a realistic perspective on the possibility of being in a survival situation will familiarize himself with the process. The centuries-old method of exploiting friction to make fire is to use a bow drill. This invaluable implement consists of four simple components – a bow, a drill, a fire block and a socket.

Bow Any pliant stick 65-85cm (26-33in) long and about 2cm (0.75in) thick can be used as the basis of the bow. It must be springy enough to tauten the bowstring so that it will grip the drill when the device is operated. Use a length of leather thong or cord to bend the bow stick by about 15cm (6in) from the straight position. There should be enough slack to enable the bowstring to be looped once around the drill.

Drill A sound, dry length of medium-hard wood is needed for making the drill. Cedar, elm, willow, balsam fir, cottonwood or cypress are examples of the density of wood most suitable. Choose a piece from which you can cut a straight shaft of 30-40cm (12-16in) in length and about 2cm (0.75in) thick. If possible, shape the shaft so that it is octagonal in section, rather than round. This will allow the bowstring a better grip. Sharpen the wood to a blunt 60° point at one end, and a sharper 30° point at the other.

Fire Block Select a piece of wood of the same hardness as used for the drill (for preference it should be of the same species) about 20cm (8in) long, 10cm (4in) wide and 2cm (0.75in) thick. At the centre of one edge, cut a V-shaped notch about 2cm (0.75in) wide at the edge of the block but opening out a little wider on the underside. The point of the notch should extend about 2.5cm (1in) towards the centre. At the point of the V, cut a small rounded hollow, such that the V's point is at its centre.

Socket The purpose of the socket is to hold the upper end of the drill during use. Any piece of fairly hard wood which will fit the hand will do. A knot from a pine or hemlock tree is ideal. In the flatter side, cut a hollow 1cm (0.4in) across and 1cm (0.4in) deep. Shape the other side to fit the palm of the hand. Lubricate the hollow with candle grease, animal fat or soap if any of these is available.

Good tinder for use with the bow drill comes in many forms, but particularly effective is a wad of fine, soft, very dry dead grass mixed with shredded cedar or birch bark. As when making any camp-fire, you should gather supplies of tinder, kindling and main fuel beforehand.

USING THE BOW DRILL

Place the block on the ground with a piece of dry bark or thin, flat wood under the notch. Rub a small amount of tinder into a loose ball and place it on the bark near the notch.

Loop the bowstring once around the drill, and fit the blunt end of the drill into the hollow in the block. Hold the block steady with the left foot. Fit the socket over the top end of the drill with your left hand (if you are left-handed, throughout the operation use your right hand where the left is indicated, and vice versa) and steady this hand (or wrist) against the left shin. Hold the bow in the right hand, with the bowstring towards the leg and the bow stick curving away. Draw the bow back and forth with the right hand, using the full length of the bow. As the drill spins in the hollow, fragments of wood are ground to powder. Heated by the friction, this powder falls through the notch and gathers on the bark below.

When a good amount of smoke is rising from the notch, check the pile of powder. As that begins to smoke too, carefully lift the block clear of the bark. Gently fan or blow the pile of powder until it glows. Then add the small ball of tinder and continue fanning and/or blowing until this ignites.

Some further tips on the use of the bow drill may prove useful. The drill must be held steady in a vertical position, and this cannot be done unless the upper hand is braced firmly against the shin. Saw as quickly

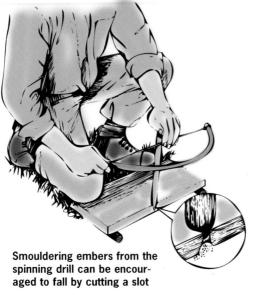

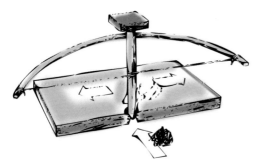

Smouldering embers from the spinning drill can be encouraged to fall by cutting a slot in the outside of the drill-hole.

Place cotton wool by the slot to catch the embers. When you see smoke, blow on the tinder until a flame appears.

as possible when smoke begins to rise. If any sand is available, a pinch sprinkled into the hollow will increase friction.

Using the bow drill is a skill which, like all others, requires practice if you are to achieve proficiency. It may take ten to fifteen minutes to produce your first glowing ember. However, as your technique improves and, perhaps, you are able to produce a better bow drill, you will reduce this time considerably. The guiding principle is: take your time and don't allow yourself to become frustrated or discouraged. It will work.

LIGHTING A FIRE

It is worth taking some care when choosing the site for your fire. If you already have a shelter, or plan to build one, you will not want it to be filled with smoke. At the same time, it is desirable to benefit from the fire's heat when you are in the shelter. It is a matter of balancing the two requirements.

Check the wind direction, and the dryness of the location. Consider the availability of fuel in the vicinity, especially if you have decided to stay there and await rescue. If conditions are windy, some form of shield from the wind must be available – either provided by nature or erected by you – before you attempt to light your fire. Although the fire will need good ventilation once it is well established, the wind can easily blow out the first tiny flames.

If you are using an open flame, hold it steady under the tinder so that the heat builds up at one point, dries the tinder if necessary, and then ignites it. As the first tinder begins to catch, add further small quantities above the new flames so that the heat builds up to an ever greater degree, so repeating the process on a larger scale.

Once a fire is burning well, it is often useful to enclose it in a circle of stones, if these are readily available. This will define the size of the fire and lessen the danger of its spreading. If larger stones are used on the windward side, the fire will be able to burn more steadily than if it is left entirely open to the wind. This is important, because a continually fanned fire consumes much more fuel than one which is sheltered, and gathering wood to feed a roaring fire can use up a great deal of time and energy.

A fire may appear to have burned out overnight. If this seems to be the case, check the ashes, for they will often retain enough heat to allow you to relight your fire. If they are still giving off warmth, gently push some tinder down into them, almost covering them. Use a twig to do this, to avoid burning your fingers. After a little while the tinder may start smoking. At this point use gentle blowing or fanning and add more tinder to relight the fire. If no smoke appears, the ashes are too near extinction to be of use.

Build a platform for the fire if you are on wet or marshy ground. Once the base has burnt through the ground will be able to support the fire.

To avoid a fire getting out of control in a forest or when it is left unattended, surround it with a ring of stones.

TYPES OF FIRES AND STOVES

Necessarily simple though most survival fires are, there are variations to suit different purposes. If you want a fire to warm yourself, you will find that it is best to make a small one. This will use less fuel than a large fire, will prove easier to control, and will not be so fierce as to make it impossible to crouch near it to gain its benefit. Some kind of reflector – a metal panel from a vehicle is ideal – behind the fire will increase the amount of heat you receive. Alternatively, you will get more heat from a number of small fires in a line alongside you, or in a pattern around you, than you will from one big fire.

PYRAMID FIRE

One type of fire well worth considering is the pyramid fire. The purpose of this is to dry out wood for future use. You may need dry wood to regenerate your fire if it has died down or has been banked down overnight. You will certainly need dry wood to prepare one of the most important types of fire – a signal fire – should you need one.

A pyramid fire has several functions. Firstly, it can form the basis of a signal fire. Secondly, it can be used to dry out timber that is very wet. Once your fuel is dry, keep it that way.

SIGNAL FIRE

Its intended purpose demands that a signal fire should be ready for instant use. Over a fresh pile of very dry tinder, together with dried or partly burned twigs, construct a pyramid of dry logs, some of which may also be partly burned.

To ensure that your signal fire will ignite quickly, you must keep it completely dry. If rain seems to be in prospect, you may have to build a shelter over the unlit fire. Clear the ground immediately around the site so that when the fire is lit, it will not spread out of control. If possible, prepare three such fires in a triangle with sides 30 metres (100 feet) long. If you have any liquid fuel – petrol, oil or similar – to spare, you can use it to speed up ignition.

If an aerosol can containing inflammable liquid is available, save it for lighting your signal fire with the minimum of delay. In extreme circumstances – for example, when you need to signal friendly forces while on the move – the aerosol alone can be used to send out a flame as a signal. Note that lighting the spray from an aerosol can is very dangerous and should be only be done when there is absolutely no alternative means of attracting rescuers' attention.

To light your signal fire with an aerosol, half-bury the can in the ground, with the jet pointing downwind towards the base of your signal fire. Have to hand a flat stone weighing about 2kg (4.4lb) and a taper at least 1 metre (40in) long. When you need to light the fire:

1 Place the flat stone on top of the can's nozzle, thus releasing the liquid jet.

2 Light the taper and hold it where the jet meets the signal fire. It must stress that while this fire-lighting method produces dramatic effects, it is highly dangerous. Each time I have tried the can has exploded.

When the blaze develops, try to produce some contrast with the background, to highlight your position as effectively as possible. On a clear day, burn vegetation to produce white smoke. If the sky is overcast, make black smoke by burning oil or rubber, if either of these is available. At night, build up the fire to generate tall, bright flames. The overriding need is to have your signal fire ready to burn as quickly and as fiercely as possible, so as to make the best of any chance of being seen by potential rescuers.

STAR FIRE

The star fire is a simple but effective arrangement. It can easily be controlled by pulling the logs outwards, which will cause the flames to die down, leaving the embers undisturbed. You can leave this type of fire unattended for up to two hours – to go hunting, for example – without the fear that it will flare up and burn down your shelter. On returning, simply push the logs inwards to bring their ends together over the embers. Fanning or blowing on the centre of the bed of embers will soon reignite them. If the weather is wet, a large flat rock placed over the charred ends of the logs will protect the fire.

IMPROVED STOVES

More economical than open fires, improvised stoves have the further advantage that a wide variety of fuels can be burned in them. Also, a stove in operation is less easily detected than an open fire from both the air and the ground, so that it is less likely to betray your presence if you are being hunted. In the desert aircraft fuels and lubricants can be burned in a stove by mixing them with sand. And, in areas where fuel is scarce, nomadic peoples have long burned animal

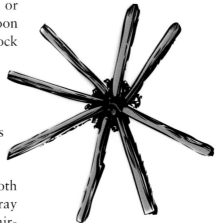

A signal fire can be quickly lit with an aerosol can. Using a stone to keep the button pressed down, light the jet only from a safe distance.

A star fire can be controlled by pushing together or pulling apart the fuel.

An oil drum makes an excellent stove. Not only does it conserve fuel, but it can also be used when it is raining.

Even in the Arctic, it is possible to find fuel to burn. A section of fatty skin, cut from a seal or other Arctic animal forms a platform on which animal bones can be burnt.

droppings in makeshift stoves.

In Arctic conditions an improvised stove can be constructed by using a wick to burn oil or animal fat. If no can is available for use in this way, it is possible to burn animal bones and fat together. This is difficult to get going and it is best to try a small version inside an igloo first. Use a candle under the fat to initially heat and ignite the fat.

By mixing petrol or kerosene with soap and sawdust it is possible to make fire-blocks which work very well in improvised stoves. They can be carried with you and used in an emergency when time is of the essence. Once they have been made they are not volatile, but it is best to store them in some kind of waterproof container.

YUKON STOVES

By far the best and safest source of fire for cooking and heating is the Yukon stove. If you are to remain in one location for more than twenty-four hours you should seriously consider building this type of stove. Rocks, stones and mud are used in its construction, which is based on the shape of a tortoiseshell. At one side you must provide a hole for the intake of fuel and air, as well as another hole at the top to act as a chimney.

Two further refinements to the Yukon stove are recommended. The first is to build a metal box or large can into the back wall, so as to provide an efficient oven. You must remember, however, that food placed in the oven will be burned unless it is separated from the metal by small sticks or stones. If twigs are used they will turn into charcoal after a day or two. Keep them for use in deodorizing boiled water if necessary, and for other medicinal purposes. The second optional improvement is to use a large flat rock as part of the top of the stove. This can be used for many cooking tasks, including griddling oatcakes, drying leaves for tea, parching grain and even frying birds' eggs.

One of the major advantages of this type of stove is that it can be left unattended while you are engaged in

FIRE-MAKING CHECKLIST

1. Choose and prepare the site for your fire.
2. Gather an ample supply of fuel and sort it into three categories: tinder, kindling and main fuel.
3. Prepare the tinder.
4. Light and build the fire – slowly. Do not smother it with kindling or main fuel.
5. Once the fire is established, make sure you derive maximum benefit from it. Begin cooking as soon as possible. Also, if you have no sleeping bag, add extra stones to the circle around the fire before you retire for the night. They will give you an extra two hours of warmth.
6. When you have cooked, dried your clothes or equipment, or warmed yourself, be sure to replenish your supply of tinder and kindling.

If you are staying in any location for some time, build a Yukon stove even though it may require carrying rocks for some distance.

other activities, and when these are done you can return to a warm fire and a hot meal. By covering the fuel and air intake with another stone you can partly control the rate of burning. In wet weather the oven enables fuel to be dried. Clothing can be laid over the outside of the stove and will dry efficiently without burning. Also, you can warm yourself without the risk of being burned.

The Yukon stove normally takes one person about two hours to construct, provided most of the materials are readily to hand. Throughout my career I have always taken the time to construct a good Yukon stove and on each occasion it has made difficult circumstances far more bearable.

If you intend using any type stove or heater inside your shelter, you must ensure that there is effective ventilation. To provide this it is necessary to make two openings in the shelter – one at the top to act as a chimney, and another close to ground level to admit fresh air. If a group are sleeping in a heated, closed shelter, one of their number should stay awake specifically to check for the presence of carbon monoxide by watching out for dizziness or nausea.

HANGI

The hangi – the name is a Maori word – is a refinement of the fire pit and uses hot stones to cook food. This is a particularly useful method of cooking while you are away from the camp, or when there are no food containers available.

Prepare a pit of a size to suit the number of survivors to be fed, bearing in mind that a hole 60cm (24in) deep and of similar diameter will accommodate enough food for a group of three or four. Set tinder and kindling on the floor of the pit. Prepare a pyramid fire of about six layers of logs, with each layer at right angles to the layer beneath. Incorporate stones into the pyramid as illustrated. Fist-sized stones are suitable, but never use soft or flaking stones, as these may well explode when heated.

If, after using the hangi, you intend to spend the night in the same spot, you can benefit from the residual warmth by making it the centre of your sleeping area. Simply put down a layer of soft bedding over the stones and the surrounding warm earth to get a good night's sleep.

Light the fire in the pit, and tend it until it ignites the log pyramid. Eventually, as the wood burns away, the heated stones will fall to the bottom of the hole. Any burning embers remaining must be removed and the ash cleared from among the hot stones. If no food containers are to hand, wrap the food in large clean leaves from a safe source. Lay the wrapped food on the hot stones, putting what needs the most cooking – almost certainly this will be meat – at the centre. Vegetables should be located further from the centre and then all other dishes introduced in order of their cooking times; for example, roots; fruit; flour cakes; leaf dishes. (Some experimentation with timing may be necessary here.) Keep all foods clear of the sides of the pit.

Cover the pit with branches and foliage strong enough to support the final covering of earth dug from

The fire in the base of a hangi pit burns through upper layers of wood interlaced with stones. After about half an hour the whole structure will collapse back into the hole.

the pit itself. This arrangement will retain the heat in the hangi, and during the next three or four hours the food will cook thoroughly. However, leaving the hangi for up to eight hours will not overcook the food.

COOKING

Water-borne diseases are among the greatest hazards

To cook, fish out any burning embers, leaving the hot stones. Cover these with edible leaves, and arrange the food on top with meat in the centre. Place more leaves over the food and fill the pit with earth. The meal will be ready after about four hours.

Survival tip
Never throw away water in which you have cooked meat or vegetables. It is full of nutrients and can form the basis of a nourishing soup. After you have boiled meat, let the fat cool on the surface of the water and, when it has set, scoop it off and use it to make a candle.

which face the survivor, and so to ensure that water is purified of any harmful organisms, it is often necessary to boil it. In addition, hot or warm drinks are a valuable source of body heat. But being able to boil water is, of course, only one of the advantages bestowed by a fire. In most cases cooking food makes it safer to eat, by destroying bacteria as well as some toxic materials and harmful products contained in animals and plants. It also makes food more palatable.

A small fire is the best way to meet the cooking requirements of a survival situation. A bed of glowing-hot coals is the ideal source of heat. An improvement on this simple set-up is an improvised stove using any available metal can or box. This is economical in terms of fuel, and is well suited to colder conditions, since, with care, it can be used inside a shelter. Such a can may be carried with you, but care must be taken.

A cooking pot can be suspended over a fire using a simple arm, made from a stick, either cantilevered or supported by a second, forked stick.

METHODS OF COOKING

Make use of a fire continually, if only to boil water. Changing the support angle alters the height of the pot to ensure slow and thorough cooking.

A fire that has reached an adequate heat and is under control can be used to cook in a variety of ways, like a domestic cooker.

Roasting This is a quick and simple way to cook reasonably tender meat. It is done by spitting meat on a stick and holding it over or near hot embers. An arm or crane, as shown on the right, is an efficient way of roasting meat. Roasting forms a crust on the outside of the meat which helps to retain the juices. If the meat is very fatty, use a drip tray to collect any surplus fat.

Even in a survival situation, vary your cooking methods. On a fire you can roast, bake, steam or boil food.

Boiling Meat that is tough needs to be cooked slowly by boiling and then finished off with roasting, baking or frying. Boiling, because it retains most of the food's value in the water, is usually the most effective cooking method. Remember, however, that the boiling point of water decreases with altitude, owing to diminishing air pressure. The higher you are, the harder it is to cook by boiling, and above 4000 metres (13,000 feet) it becomes impractical.

A section of bamboo, cut just below each of two joints, will provide a container for boiling water, as will half a green coconut. Both containers will remain unburned until after the water has boiled. Containers made from large single leaves (banana, for example) or birch bark will also hold water for boiling if they are kept moist and the fire is kept low. A square or rectangle can be folded to produce a watertight container.

Baking To bake food you must enclose it and then subject it to constant medium heat. The enclosure may be either an oven, a pit beneath a fire, any closed container, or even a wrapping of leaves or clay. Any of these is best used with glowing embers.

Steaming If you have no cooking container, nor any means of making one, and wish to prepare food that requires little or minimum cooking, you might try

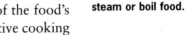

TIPS ON COOKING MEAT

— Animals larger than the domestic cat should be boiled before roasting or baking. Larger animals will need to be cut into manageable pieces before boiling.

— Roasting, unlike boiling, should be done as quickly as possible, since a slow fire's direct heat will toughen the meat.

— Meat that is very tough is best stewed with vegetables.

steaming it. Prepare a pit by putting in a thick layer of heated stones, cover these with leaves and then place the food on top. Use additional leaves to cover the food, and push a stick down into the food space. Pack an earth layer over the leaves and around the stick. Now withdraw the stick, leaving a hole that goes down into the food space. Pour water into the hole. When it heats up and turns into steam your food will be cooked

Food can be steamed in an adapted hangi pit. Adding water every ten minutes through a hole will cook food in roughly two hours.

9

A SURVIVAL DIET

In a survival situation you should devote a substantial part of your energy to obtaining at least one hot meal a day. Some foodstuffs, in particular nuts and soft fruits, are suitable for eating raw. But in other cases, particularly meats and vegetables, it is best to prepare and cook your food as this will help make it safe, more palatable and more digestible.

PREPARING FOOD

The three following chapters discuss the main kinds of food, as well as the range of cooking techniques, that are available to the survivor. This chapter looks at the preparation of food from the same wide variety of sources and at methods of preserving foods. Before they can be cooked, most forms of food, whether from animals, fish or plants, require some preparation, the most important elements of which are washing, cleaning, skinning, and plucking or scaling.

A bird is killed by pulling its neck. When you are plucking – whlch Is most easily carried out in warm water – retain large feathers for use as arrow flights. Uneaten innards make good fishing bait.

BIRDS

Generally birds should be plucked and cleaned before they are cooked. The skin should be left in place as it will help to retain the food value. It is easier to pluck a fowl after plunging it into boiling water, although waterfowl are best plucked dry. After plucking, cut the bird's head and feet off, then make a small opening in the stomach. Draw the insides out through this hole. Wash the carcass thoroughly, inside and outside, with fresh, clean water. The neck, liver and heart can be used as the main ingredients of a stew.

If time is short, it is quicker to skin a bird than it is to pluck it. Skinning is best carried out after gutting and cleaning the carcass. Small birds, after they have been gutted and cleaned, can be baked in a clay case.

It is important to note that all scavengers – vultures, buzzards, carrion crow and similar birds – must, before they are roasted, baked or grilled, first be boiled for at least 20 minutes to kill any parasites that may be present. Only then should you use other these other methods of cooking. It is a good idea to retain clean feathers for use as insulation in clothing or bedding.

FISH

As soon as you catch a fish, bleed it by cutting out the gills and other large blood vessels, which will be found next to the backbone. Scrape off the scales. (Some fish, including catfish and sturgeon, have no scales, and should be skinned.) Gut the fish by cutting open its stomach and scraping it clean, washing out any remaining particles.

A fish of less than 10cm (4in) in length does not require gutting, but may need scaling or skinning. Remove the fish's head, unless you intend to cook the whole fish on a spit. Fish may be roasted by direct heat, baked using an improvised oven or other wrapping, or

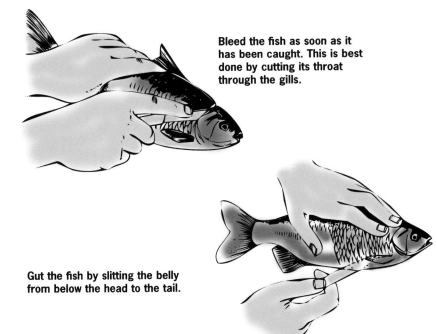

Bleed the fish as soon as it has been caught. This is best done by cutting its throat through the gills.

Gut the fish by slitting the belly from below the head to the tail.

fried. Note that you should never eat fish raw if any means of cooking it is available to you.

ANIMALS

The sooner you can skin and dress a carcass after an animal's death, the easier this task will be. Small and medium-sized animals should be hung head downwards with their throats cut to allow the blood to drain into any available container. Once it has been boiled thoroughly, the blood, which is valuable for its salts and food value, can be drunk.

Cut the skin right around the animal's knee and elbow joints. Continue down the front of each hind leg, joining the cuts and extending down the belly in a Y pattern. Cut from the belly to each elbow joint. Make a complete circular cut around the sex organs. Then, from each knee downwards, start peeling back the skin until it is removed entirely.

Next, cut open the belly, using wooden skewers made from twigs to pin back the flesh. Remove the guts from the windpipe upwards. Clear the entire mass of guts

Cut around the ankles and down the inside of the thigh towards the genitals. Next cut the animal straight down the belly as far as the throat. It should be possible to skin the animal. With the carcass suspended, open the belly and let the guts fall out. Remove all edible parts: heart, liver and kidneys. Smaller animals, such as rabbits, are also easier to skin and gut if hung.

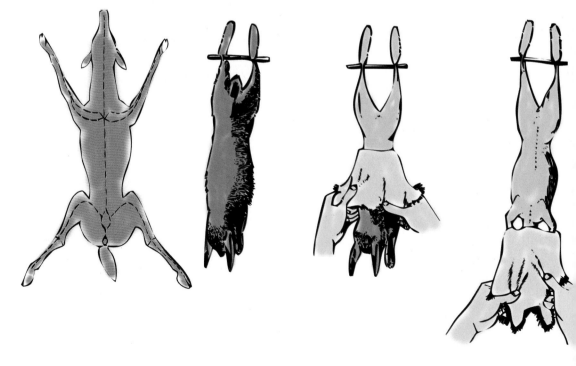

and then use a deep, circular sweep of the knife to remove the sex organs cleanly.

Throw away the entrails, and the glands in the anal and reproductive area (unless they are to be used as fishing bait). Make use of the kidneys, liver and heart, and the fat around the intestines. The meaty parts of the skull are all edible, including the brain, tongue and eyes. Keep the skin, to clean and dry for later use – for example, as crude shoes (*see* page 62).

Larger animals are dealt with in the same way, although their size and weight may be too great to allow you to hang the carcass and you may have to spread it out on leaves or grass to work on it.

COOKING SMALL MAMMALS

Small mammals and birds may be cooked either whole or cut into parts. They can be roasted, baked or boiled. However, smaller game is best boiled in order to minimize the shrinkage that tends to occur with the two other cooking methods. All entrails and reproductive organs must

It is easier to gut and skin larger game if you hang it from a tree.

A quick method of skinning and gutting a rabbit is to cut a half-moon-shaped circle, penetrating both the fur and gut. Cut off the rear legs at the second join. Raise the rabbit above your head and swing down between your legs. The guts will be cleared.

Hedgehogs are normally covered in fleas as well as sharp spines. Kill the animal by cutting it under the chin, then roll it in a thick layer of mud. Place the ball on the fire for about two hours then remove it and allow it to cool. Remove the clay and you will find that the spines and skin come away, leaving the cooked flesh exposed. Hedgehogs make excellent eating, and are very rich sources of protein.

be fully removed before the animal is cooked. Carcasses, especially those of birds, can be made more appetizing by stuffing them with edible berries, grains, roots, greens or nuts.

Rabbits can be very good to eat, although they lack the fat which provides the survivor with a vital reserve of energy. To skin a rabbit, make a cut behind the head that is large enough to allow two fingers to be inserted. Peel the skin back, and sever and remove the head and lower limbs. Cleaning is carried out by cutting down the belly and opening the body. A sharp shake will cause most of the intestines to fall out. Remaining pieces of gut should be scraped away and the cavity washed out thoroughly.

Rats and mice must be skinned, gutted and washed, and then boiled for about ten minutes before being added to a stew. Other edible animals include dogs, cats, hedgehogs, porcupines and badgers. All of these must be skinned and gutted before they are cooked, and, like rats and mice, they taste best when they are cooked in a stew which includes dandelion or other edible leaves.

Another good option for cooking small animals is to bake them in a casing of clay in the time-honoured manner of the Romanies. Carcasses cooked in this way must be gutted and cleaned, but the clay, when it is removed after cooking, will take with it the hedgehog's spines or the small bird's feathers, revealing flesh cooked and ready for eating.

REPTILES

Many reptiles, including snakes, lizards, frogs, alligators and turtles, are edible. The head and skin should be removed before they are cooked. The meat of snakes is delicious, but it is not always a good idea to go chasing them. Small snakes, lizards and frogs can be spitted on a stick and roasted. Larger eels and snakes are better if they are boiled before roasting.

Turtles should be boiled until the shell comes off. The meat can then be cut up and used to make a soup with any edible plants that are available. Salamanders are edible if they are spit-roasted.

SHELLFISH

An excellent basis for a soup also including green vegetables, shellfish may be boiled, steamed or baked in the shell. Crayfish, crabs, shrimps and prawns must likewise be cooked so as to destroy the organisms they contain, as these are capable of inducing diseases in humans. Cook them alive by dropping them into boiling water. You should do this at the earliest moment possible after catching them, otherwise they will spoil very quickly.

PLANTS AND VEGETATION

Some plants and leaves can be eaten after they have been briefly washed in fresh, clean water. However, in general it is best to cook all food plants. The same rule applies to berries. Once cooked, both plants and berries are best added to other dishes, so as to add flavour and enrich the meal's nutritional value.

Roots and tubers After they have been washed, roots and tubers are best baked or roasted, although they can be boiled.

Herbs Leaves, stems and buds of herbs must be boiled until they are tender. Any bitter taste can be diminished by replacing the water several times.

Nuts, grains and seeds Most nuts, grains and seeds can be eaten raw, but many are better if they are parched. Parching means slow heating until the food is well scorched. This is best done in a metal container, but can be achieved on a hot flat stone – the top of a Yukon stove is ideal. Some nuts – for example, acorns – are better crushed than parched. Chestnuts are excellent whether roasted, steamed or baked, and they can also be eaten raw.

Fruit Many fruits are a valuable food when eaten raw. However, any which are tough or have heavy skins should be baked or roasted, while succulent fruits can be boiled.

Survival tip
All frogs and snakes must be skinned before they are cooked because in some species the skins contain poison.

Eggs One of the most convenient and safe foods, birds' eggs may be eaten even during the development of the embryo. If they have been hard-boiled, they will provide a convenient and easily portable food reserve which will keep for several days.

PRESERVING FOOD

Rare is the survivor who can count on a regular and continuing supply of wild food. A couple of days' good hunting or gathering may be succeeded by a period when food is harder to find, catch or trap. The weather may make food gathering, and indeed any other outside activity, difficult. It is important, therefore, to know how to preserve food. Doing so will allow regular supplies to be maintained or, if you plan to move, a portable reserve to be built up. You must do all you can to avoid food wastage through deterioration, which is a particular problem in hot climates.

Both fish and red meat can be preserved by smoking. Build a tepee frame with a food platform halfway up. Build your fire and, once established, cover it with green leaves. You will get better and quicker results if you cover your tepee to contain the smoke.

SMOKING

In temperate climates or during the summer, meat, game and fish which are to used soon should be stored in the coolest, shadiest place available. However, food

Food can also be preserved by air-drying. Although you can hang the flesh from the branches of a tree, it is better to build a frame.

which is surplus to your immediate requirements can be preserved by drying. This can be done naturally, in the sun and the wind, but is probably better achieved by using the smoke from a fire, as the North American Indians did with the smoke tepee.

All meat to be smoked should be cut into thin slices, approximately 1cm (0.4in) thick, across the grain.

These should be laid on the smoking platform and left in the smoke until they are brittle. Do not use wood which

FREEZING

In cold climates surplus food can be easily and quickly preserved by freezing. Cut it into strips or small pieces and spread it on the ground outside your shelter. Guard it against animals while it is freezing. When it is frozen, store it safely above ground level – at a minimum height of 2 metres (6.5 feet).

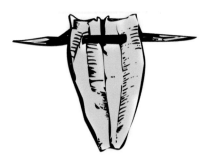

Skinned fish can be dried most efficiently if, once it has been split open, it is held flat with a wooden skewer.

contains pitch (pine and fir are the most common examples in the northern hemisphere) as its acrid smoke can taint the meat. Once your fire is going, it is essential to keep adding green wood or foliage to make smoke. If you have a tent or tepee, this can be used for smoking food, but take care to avoid damage to the shelter. If there are any ventilation flaps they should be kept closed while smoking is in progress.

Fish are prepared for smoking by removing the heads and tails so that they can be spread flat. Thin twigs can be used as skewers to hold them so.

Plants can also be dried by leaving them in the sun or by smoking. Fruits are best cut into thin slices to speed drying in the sun. Berries can be converted into jam. Mushrooms dry easily and keep well, but once dried they should be soaked in water for at least half an hour before use.

10

PLANTS FROM THE WILD

Obtaining sustenance from wild plants is a survival skill well worth acquiring – not just because of their abundance in the kinds of environments in which survival situations arise, but because they may be the main, or even the only, form of food available. However, less than half of all wild plants are edible, and most of them only in part. Detailed knowledge is therefore needed if you are to take advantage of nature's bounty. If you know which plants to look for, and which to avoid, you should normally be able to find enough food to keep yourself alive.

EDIBILITY TEST

If you are not certain of a plant's identity, or need to establish if a plant is edible, follow the Edibility Test carefully:

1   Never collect plants from polluted waters or other contaminated areas. Always clean all plants thoroughly before attempting to eat or cook them, and remove damaged or inferior parts.

2   Do not assume that every part of a plant is edible because you have found that any one part is.

3   Do not waste time and effort testing any plant unless it is abundant and easily obtainable.

4   Test for the presence in the plant of any contact poison. Crush the leaf and rub some sap on to the inside of the wrist. Wait for fifteen minutes. Then, if no itching, blistering or burning occurs, go on to step 5.

5   Hold a small portion of the plant in the mouth for five minutes. If no unpleasant reactions occur, chew the plant, again looking for unpleasant signs (extreme bitterness, burning or soapy taste). If there

are none, swallow the juice, but spit out the pulp. Wait for eight hours.

6 If no ill effects develop – for example, nausea, dizziness, sleepiness, stomach aches or cramps – eat a teaspoonful of the plant and watch for these effects for a further eight hours.

7 If no negative effects occur, eat about a handful of the plant. A final twenty-four hours without trouble indicates that the plant is safe and can be eaten in larger quantities.

8 Eat only healthy plants, avoiding all with rotting parts, mould, diseases or insect infestation.

9 Avoid any plants with milky sap (except dandelion, goat's beard and coconut) or with caustic sap or a bitter or burning taste. Always boil leaves which have prickly hairs (for example, stinging nettle).

10 Test only one plant at a time, and on only one person at a time, so that the cause of any ill effects can be pinpointed.

11 Even palatable wild plants may prove detrimental to health if eaten in large quantities or in smaller quantities over a long period. Wherever possible, make a salad or vegetable stew, combining leaves, berries, nuts, inner bark and rootstocks (rhizomes). You will achieve a more balanced diet, as well as a tastier meal. This is one occasion when variety really is the spice of life.

12 The test does not apply to fungi.

There follows a catalogue of plant forms found in the wild of which the survivor should be aware – because they are a potential source of sustenance, and, in some cases, because they are also of medicinal value. Others are listed because they are commonly encountered but are poisonous. The list comprises six sections: PLANTS; FRUITS; ROOTS, ROOTSTOCKS AND TUBERS; NUTS; SEAWEEDS; FUNGI.

> **Survival tip**
> The Edibility Test may seem to be time-consuming and overcautious, but one of the basic strands of any survival technique is to choose the safest option. Start testing available plants before food stocks are exhausted. Even better, practise edibility testing as a survival technique as part of your general preparation for outdoor activities. Be sure to keep notes of your experiments.

PLANTS

This section lists some of the most useful of the food plants that are readily available in the wild. In many cases they have a wide distribution in the north temperate regions.

GREAT PLANTAIN *(Plantago major)*

A slightly hairy perennial widespread throughout the temperate regions of the northern hemisphere, the great plantain has large oval leaves forming a loose basal rosette. The stem, 20-45cm (8-18in) in length, carries very small flowers in a closely packed spike. Young leaves can be eaten raw and older leaves, with the fibrous ribs removed, cooked as greens, though they remain rather bitter. The latter are best used as an ingredient of stew.

> ✚ *An infusion made from the dried leaves (put two teaspoonfuls in a cup of boiling water and allow to stand for ten minutes) can be used for bronchitis and coughs.*

DANDELION *(Taraxacum officinale)*

This perennial is widespread throughout the north temperate regions. Its leaves, shiny, bottle-green and deeply toothed, are some 15cm (6in) long and form a rosette at the plant's base. The inner leaves are erect, the outer leaves shorter and spreading. The flowers are yellow and solitary, and appear on stems up to 30cm (12in) long between March and August. The dandelion produces copious milky sap. Young leaves may be eaten raw – their bitter taste is alleviated if they are soaked for two hours in cold water. Developing shoots, apparent before the stems begin to grow, can be used like Brussels sprouts. Older leaves, with the rough central vein removed, are best boiled. Cleaned roots can be boiled like potatoes, and have a pleasant taste. Sun-dried roots, baked and crushed, provide a coffee substitute. The dandelion is a valuable source of food for the survivor, particularly because its leaves and roots are

Dandelion

available throughout the year.

✚ *Chop and boil fresh leaves to produce a strong-tasting liquid. Take one spoonful of this for stomach complaints.*

BRACKEN *(Pteridium aquilinum)*

Widely distributed throughout the world, this perennial has large fronds 20-200cm (8-80in) long (but in some cases up to 400cm/13 feet) growing singly from a base densely covered with short, soft, rust-brown hairs. Young fronds are coiled inwards, and become three-pronged as they unfold, while older fronds are more clearly three-pronged. Gather uncoiling young fronds, brush scales or fine hairs from the stalks, and wash and boil until tender (approximately thirty minutes). The underground stems can be roasted and the inner portion eaten. Avoid male bracken, which has single fronds and is sometimes found in similar habitats.

STINGING NETTLE *(Urtica diotica)*

This perennial herb is distributed throughout the temperate regions of the world. Its stems grow up to 120cm (47in) in height and carry heart-shaped, toothed leaves 3-8cm (1-3in) long. These are covered with fine hairs which irritate the flesh when they are touched. The stinging nettle is found in woodland and sheltered, grassy places, often in large colonies. Young shoots gathered in March or April, before flowering, can be eaten fresh, provided they are dipped in boiling water to remove formic acid from the 'stinging' hairs. They are a very good source of Vitamin C. Other leaves should be chopped and boiled to remove acid, but boiling should last no longer than six minutes, in order to retain the maximum food value. Nettle leaves can also be used as an ingredient for stew, and, when they are dried and rubbed into flakes they make very acceptable tea.

✚ *Freshly pressed nettle juice (one or two teaspoonfuls per day) makes an excellent pick-me-up after an exhausting day.*

Bracken

Stinging nettle

CLOVER – red *(Trifolium pratense)* and white *(Trifolium repens)*

These similar perennials are widely distributed in the north temperate regions, especially in grassy places. Red clover is slightly more erect than white, but the stems of both are 12-60cm (5-24in) in length. The trefoil leaves are 1-5cm (0.2-2in) long. Both clovers flower between May and September. Young leaves can be eaten raw; all leaves can be cooked or used in stews. The dried flowers make a fine tea.

WATERCRESS *(Nasturtium officinale)*

A perennial occurring throughout the north temperate regions, watercress also flourishes in some areas in the southern hemisphere. It is commonly found in running water, but usually in small streams rather than large rivers. The flowers, which appear between May and October, are 5mm (0.2in) in diameter and form small groups. The leaves, which remain green in autumn, grow on stalks from angular hollow stems which are up to 65cm (26in) long. Up to ten leaflets grow in pairs on a stalk, at the end of which there is a larger, heart-shaped leaf. Leaves and stalks can be eaten raw. Take care to remove snails and insect leaves. Boil the plant if you have any reason to suspect water pollution.

COMFREY *(Symphytum officinale)*

This erect hairy perennial herb, 25-120cm (10-47in) tall, is found in damp localities and on the edges of woodland and copses in all temperate regions. The leaves are rough and shaped like spearheads. The flowers, which appear between May and June, are funnel-shaped, 2cm (0.8in) long, nodding and pale-yellowish, white or purple. Young shoots can be eaten raw, or with young leaves cooked or added to stews.

SORREL *(Rumex acetosa)*

A perennial which occurs throughout the north temperate regions in grassy areas and woodland clearings, sorrel has smooth, erect, branching stems of 20-70cm

Red clover and white clover

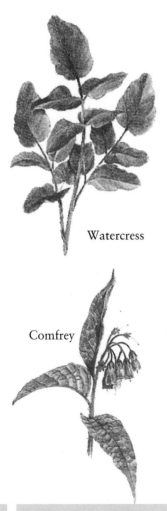

Watercress

Comfrey

(8-28in) in length. It produces red flowers, grouped on stem branches, between May and July. The leaves, which are 3-12cm (1-5in) long, grow from the base on long stalks. Young shoots and leaves, rich in vitamins, can be eaten raw, and all leaves are usable in soups. Consume this plant in moderation.

✠ *Place fresh leaves in a clean piece of cloth and gather this into a ball. Moisten the ball by dipping it in boiling water, then beat it with a stick to crush the leaves. Squeeze the ball so that liquid filters through the cloth. Use it to treat open cuts and skin eruptions.*

Sorrel

SEA BUCKTHORN *(Hippophae rhamnoides)*
This tall, heavily branched shrub grows to 0.8-3 metres (31in-10 feet) in height and has long, thin, almost sessile, leaves covered with minute scales. The flowers, 3mm (0.1in) across, form a small receptacle shape, while the fruit, 5-8mm (0.2-0.3in) in diameter, is bright orange. Sea buckthorn is found along sea cliffs and further back from the water, among sand dunes. The freshly pressed juice is used as a preservative when mixed with honey and other fruit. It can also be used to sweeten herbal tea.

✠ *Freshly pressed juice is an excellent remedy for the common cold, tiredness and exhaustion.*

Sea buckthorn

Parsley

PARSLEY *(Petroselinum crispum)*
A biennial herb found on rocky wasteland, parsley has lower leaves that sprout directly from the tap root, while the upper leaf is spear-shaped and forms small, crisp clusters. This well-known herb is used to flavour food, but the individual small leaves can be eaten as a salad. Parsley and other fresh herbs are rich in vitamins.

DAISY *(Bellis perennis)*
This perennial herb is widely distributed in grassy places. Its leaves, which form a basal rosette, are paddle-shaped and 2-9cm (0.8-3.5in) long. The flowers are white, 2cm (0.8in) in diameter, and are solitary on

Daisy

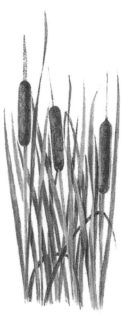

Reedmace

an erect stem 40-100cm (16-39in) tall. Young flower buds and young leaves can be eaten raw in salads or added to soups. The daisy makes good survival food because it is available all year round.

> ✛ *An infusion made by soaking the leaves in cold water for ten minutes is a good remedy for diarrhoea.*

REEDMACE *(Typha latifolia)*

An aquatic herb found throughout the world, except in the extreme north and south, reedmace (also known as cat's tail) grows on the banks of rivers, ponds and lakes. In slow-moving water it is often the dominant plant. The erect round stems reach 1-4 metres (39in-13 feet) in height, while the erect leaves grow to 2 metres (6.5 feet) in length and 2-6cm (0.8-2.4in) in width, sheathing at the base. The flower is a very dark brown, sausage-like spike 15-30cm (6-12in) long. Young leaf shoots are edible after they have been boiled for three to eight minutes. The yellow pollen can be used to make bread. The rootstocks are rich in starch and sugar, and are edible boiled or raw. Remove the outer covering of these, then grate or chop the inner white parts.

GOAT'S BEARD *(Tragopogon pratensis)*

Annual to perennial, this herb is found in all but the northern extremities of the north temperate regions. Its stem, 20-70cm (8-28in) tall, grows from a half-sheathing base which itself tapers to a long point and is clearly white-veined. The upper leaves clasp the stem. Goat's beard flowers between June and July, producing solitary yellow flowers on long stalks which are thickened at the top. The seed-head is hairy (rather like a dandelion 'clock'). The stems with young buds can be treated in the same way as asparagus: the young leaves, tips of shoots and dried tap root can be used as salad ingredients. The whole plant can be used, and is best cooked in a stew or soup.

WATER PLANTAIN *(Allisma plantago aquatica)*
This erect aquatic or semi-aquatic perennial, growing up to 1 metre (39in) tall, occurs throughout the north temperate regions. Its few leaves are almost heart-shaped, 15-20cm (6-8in) long, and on long stalks. The numerous 1cm (0.4in) flowers are either white or pink. The flower stems are 30-120cm (12-47in) long and form a loose pyramid. The thick rootstocks below ground lose their bitter taste when dried; they are cooked like potatoes.

Water plantain

HAWTHORN *(Crataegus monogyna)*
A deciduous shrub or tree reaching up to 6 metres (20 feet) in height, the thorny, much-branched hawthorn (also known as may) is found in woods, hedges and scrub in the north temperate regions. The flowers, which are seen between April and May, are small, white and bunched. The fruits, red and 1cm (0.4in) in diameter, appear between July and October. Young shoots can be eaten raw, while the fleshy fruits are edible and not bitter.

ROWAN *(Sorbus aucuparia)*
Also known as the mountain ash, the rowan is a decid-uous slender tree which grows to 20 metres (66 feet) in woodland and mountain areas throughout the north temperate regions. The leaves are up to 26cm (10in) long, and comprise up to nine pairs of dark-green leaflets. The 1cm (0.4in) flowers, which appear between May and June, are white and form clusters. Also 1cm (0.4in) in diameter and clustered, the fruits are red with yellow flesh. The latter are edible and good for use in soups. Boil them briefly, and discard the water to remove bitterness.

Rowan

FRUITS

Edible fruits should not be overlooked as a food source, since, in season, they can be plentiful. Do not be tempted to overeat if a good supply is found. Like most wild plants, they will make you sick if you eat too many at one time. Take the opportunity to collect and dry as many as possible against future needs. They are produced in all but the most extreme of the climatic zones.

BLACKBERRY *(Rubus fruticosus)*

Blackberry

This perennial deciduous bramble is widely distributed throughout the north temperate regions, in woodland, hedges, heath and scrub. It is easily identified by its long stems, armed with thorns of various kinds, which wander and intertwine, often forming large clumps. The pink or white flowers appear between June and August. The fruits are first green, then red, and finally shining black. These berries can be eaten raw, or collected and used for drinks and in salads. The leaves can be dried – this should be done very slowly – and crushed, to make very good tea. They can also be used for tea if very fresh. The related dewberry *(Rubus caesius)* is found in similar areas and produces a similar fruit with fewer but larger segments. Warning: Do not use wilted leaves from the blackberry, dewberry, raspberry, peach, plum or cherry. All can be poisonous when in this condition.

DOG ROSE *(Rosa canina)*

Dog Rose

A deciduous shrub up to 4 metres (13 feet) tall, the dog rose is found throughout the north temperate regions in woodland, hedges, scrub and rough hillsides. The stems are erect or arching with stout curved thorns. The leaflets are oval, toothed and up to 4cm (1.6in) long. The flowers, which appear between June and July, are 5cm (2in) in diameter, and have five pink or white petals. The scarlet fruits, rose-hips, are egg-shaped, 2cm (0.8in) long, smooth and shining. Gather these, a good source of Vitamin C, between August and December.

Hips must be cut open and the seeds and fine hairs removed before they are eaten raw or cooked in a pie or soup.

BILBERRY *(Vaccinium myrtillus)*

As far north as the Tundra (Asian, American and European), bilberries (also known as blueberries, whortleberries or hurtleberries) can be found. In slightly milder conditions they grow with heather. Further north, or higher on mountainsides, they replace it completely, usually growing only 15-45cm (6-18in) high. Further south they may grow to 2 metres (6.5 feet). The berries, which are dark blue or black, can be eaten raw, although they are less bitter when boiled.

Bilberry

CLOUDBERRY *(Rubus chaemorus)*

Another inhabitant of the regions in which the bilberry occurs is the cloudberry. Likewise occurring in large colonies in some places, it grows to 10-20cm (4-8in) has simple leaves and carries an orange-red fruit at the plant top. The cranberry and the cowberry (both with red berries) can be found in similar areas, but are less abundant. Their fruit is rather acid.

MULBERRY *(Morus nigra)*

Found in warmer areas in the temperate regions, the mulberry is a dense shrub or tree reaching up to 15 metres (49 feet), with stout branches and dark-brown, scaly bark. The fruit, up to 4cm (1.6in) long, is blackberry-like but dark purple in colour, and appears in late summer or autumn.

A relation of the mulberry – the wild fig – is found in the tropics and the subtropics. The fruit of its many varieties grow directly out of the branch wood. Figs which are edible will be soft, and coloured black, green or red. Another mulberry relative is the breadfruit – a tree common in the tropics. Up to 13 metres (43 feet) tall, it has tough smooth leaves 30-90cm (12-35in) long. The fruit can be eaten raw or cooked in the embers of a fire.

Cloudberry

Mulberry

Crab-apple

Wild grape

CRAB-APPLE *(Malus sylvestris)*
Common in the same regions as the mulberry, the crab-apple, a shrub-like tree with spiny, tangled branches, is found in woods, exposed heaths, field and hedges. The apples, which are ripe in late autumn, can be sliced and dried as part of the food store.

WILD GRAPE (Genus *Vitis)*
The wild grape vine grows in both warm temperate and subtropical areas, and is widely distributed. Its fruit is rich in sugars.

WILD STRAWBERRY *(Fragaria vesca)*
A perennial, small, rambling herb found in woodland and grassland, the wild strawberry produces a fruit which is sweet and resembles that of the cultivated variety but is much smaller. It is rich in vitamin C.

✢ *Fresh leaves boiled together with the fruit make a refreshing tea drink which is beneficial both for the digestion and an upset stomach.*

Wild strawberry

BIRD CHERRY *(Prunus padus)*
This deciduous, spreading tree grows to 3-10 metres (10-33 feet) in height, and has a peeling bark with an unpleasant smell. It is found along the banks of small rivers or on wet open pasture land. The leaves are flat and oval-shaped and the flower has five white petals with yellow centres. The fruit of the bird cherry is small, black and very shiny.

✚ *This tree has no edible parts but it is worth mentioning for its medical value. An infusion made by boiling fresh bark (when the tree is in flower) makes an excellent sedative and is also an effective painkiller.*

Bird cherry

ROOTS, ROOTSTOCKS AND TUBERS

A wide variety of plants that can be a source of food for the survivor lie beneath the ground in the form of roots, rootstocks and tubers.

TI PLANT *(Cordyline fruticosa)*
Occurring in both cultivated and wild forms, and growing to 2-5 metres (6.5-16.5 feet) tall, the ti plant (also known as the good-luck plant) has a wide distribution. It has clusters of broad, shiny leathery green (occasionally red) leaves at the ends of thick stems, and red berries. The rootstock, a good source of starch, is best baked.

FLOWERING RUSH *(Butomus umbellatus)*
Found at the edges of ponds, lakes, on river banks and in marshy places, often in shallow water, the flowering rush grows to 1 metre (39in) and produces loosely clustered green and pink flowers. The rootstocks below ground are edible if peeled and boiled.

TARO *(Colocasia esculenta)*
Moist forested areas of the tropics are the habitat of the taro. The single stem, reaching up to 1.6 metres (5.3 feet), carries large leaves up to 60cm (24in) and roughly

Ti plant

Wild potato

heart-shaped. The yellow flower, 40cm (16in) long, grows from a white, leaf-like organ. Occurring just below ground, the tuber is edible but must be boiled, to destroy harmful constituents.

WILD POTATO
The tubers of this plant, found throughout the world, are edible. However, the plant itself is poisonous.

SOLOMON'S SEAL *(Polygonatum multiflorum)*
If given lengthy boiling, the tubers from this small, broad-leafed plant are edible. Afterwards the water should be discarded. The plant's name derives from a legend which says that King Solomon endorsed the use of the powdered root for bruises. It may indeed prove useful for this purpose. Warning: The berries of this plant are poisonous.

WATER CHESTNUT *(Trapa natans)*
This non-rooted aquatic is found in large colonies on quiet water. It has sub-aquatic leaves which are long

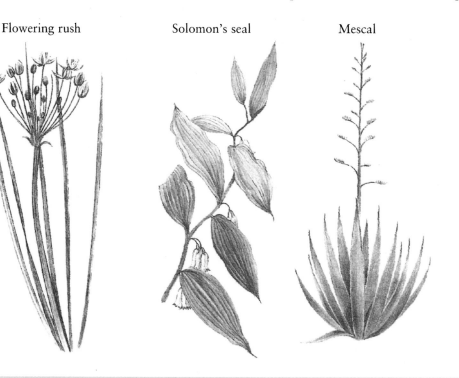

Flowering rush   Solomon's seal   Mescal

and feathery, as well as floating leaves in a surface rosette. White flowers appear in summer. Nuts, carried below the water surface in 'horned' pods, are edible.

MESCAL

Typically a desert plant, this cactus also grows in most tropical areas. It has clustered, erect leaves which are tough, thick and sharply pointed, and a stalk which rises centrally, carrying yellow flowers in a loose spike. The shoot is edible before the flowers are fully developed, and is best roasted.

BAMBOO

Found in moist areas of forest and river and stream banks, bamboo stems grow to 5-25 metres (16.5-82 feet). Young bamboo shoots are edible, and should be boiled with one or two changes of water to remove the bitter taste. The seeds of the bamboo flower are edible if boiled like rice.

WILD DESERT GOURD

This creeping plant, up to 4 metres (13 feet) long, produces edible flowers and an orange-sized gourd containing seeds which are edible if boiled or roasted. The stem shoots can be chewed to extract water.

Bamboo

Wild desert gourd

Wild rhubarb

Arrowhead

Baobab

WILD RHUBARB
Found on mountain slopes in open aspects on edges of woods and stream banks, wild rhubarb has long, stout stalks with large, green leaves growing from their bases. The stem is edible after boiling to alleviate its strong, bitter taste.

ARROWHEAD *(Sagittaria latifolia)*
A large aquatic plant with spear-like leaves, the arrowhead thrives near fresh water. The flowers are distinct by their three petals, pale green to yellow in colour, and their black seed centre. The tubers are edible raw, but the flavour is improved by boiling or roasting.

PRICKLY PEAR *(Opuntia ficus-indica)*
Occurring in many desert areas, the prickly pear has a thick stem, 2-3cm (0.8-1in) in width, which holds water. Carried on the stem are clusters of pads covered with sharp prickles. Yellow or red flowers grow at the top of the upper pads, producing egg-shaped fruit up to 10cm (4in) long. These are edible if opened and the outer layer removed, so that the pulp and seeds inside can be removed. The inner portions of the pads can also be eaten, either raw or boiled, but for cooking are best cut into strips. This plant does not produce any milky juice. Avoid any similar plant that does.

BAOBAB *(Adansonia digitata)*
Found in open tropical bush, particularly in Africa, the baobab is distinguished by a trunk the width of which is half the tree's height. The leaves are edible when used in soups, and may also provide water. The large, pulpy, pendulous fruit, which follows large, white flowers, is also edible.

NUTS

Among the most valuable of all plant foods are nuts, because they provide fats, vitamins and protein. Plants producing edible nuts occur in all continents and in every type of climate except the polar regions. Nuts found in the tropics include coconuts, brazil nuts (or shoenuts) and cashews. The temperate regions produce hazelnuts, beechnuts, sweet chestnuts, walnuts, acorns, almonds and pine nuts among others. Wild pistachio nuts grow in some desert and semi-desert areas in parts of East Asia.

Most of the nuts listed above may be eaten raw, or chopped or ground for adding to soups. Some nuts – acorns, for example – are bitter. Their taste can be improved if they are boiled for two hours and then soaked for three or four days in cold water. They can be ground into paste, which can either be dried to make flour, or used with water as the basis of stew or soup, or on its own as a gruel.

Wherever nuts are available, gather and store as large a quantity as you can. Once shelled, they provide a rich and easily portable source of food. They see squirrels through the winter – and they can keep the survivor going too.

Hazelnut

Pistachio

Sweet chestnut   Pecan   Wild chestnut

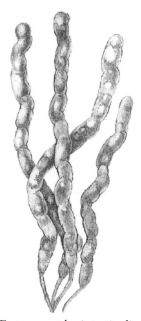

Enteromorpha intestinalis

Bladder wrack

SEAWEEDS

Any survivor on a seashore is almost certainly within reach of a wild-plant food source – seaweeds. These are to be found between the high-tide and low-tide lines, and the most plentiful growth occurs in shallow water just below the low-tide line. This is because seaweeds grow by absorbing their nutrients directly through their fronds from seawater.

When you are gathering seaweed to eat, look for fresh plants – those that are either still attached to rocks or other features, or floating freely in the water – since many types of seaweed spoil quickly if they are left exposed to the air. Always wash the crop in clean water, using as many changes of water as are necessary to remove any sand or minute crustaceans. Green, brown and red species of seaweed provide good food and all are good sources of Vitamin C and iodine. Seaweed dries best when hung in a dry, well-ventilated area for two to three days.

ENTEROMORPHA INTESTINALIS
This widely distributed seaweed has no common name in English, but the second half of the Latin name assists identification. Green, tubular and with a diameter of 1–3cm (0.4–1.2in), it is constricted at intervals so that it resembles an intestine. It reaches 5–60cm (2–24in) in length and is found grouped on rocks or in rock pools near the high–tide line. Eat this seaweed raw, or dry it for use in soups.

BLADDER WRACK *(Fucus vesiculosus)*
Found on rocky or stony north temperate coasts, the bladder wrack occupies a distinct zone at mid-tide level. The fronds are olive-brown to dark greenish-yellow, thick and leathery, and grow to 15–90cm (6–35in) long. The stalks have a distinct midrib, and the fronds carry paired air bladders which cause the plant to float up towards the light. Fresh or dried fronds can be used, boiled in soups or stews.

SEA LETTUCE *(Ulva lactuta)*

The coasts of both the Atlantic and the Pacific are the habitat of this seaweed, which is found among rocks and in pools, usually between tide lines. Lettuce-like in appearance, it ranges in colour from light to dark green. It can be eaten raw, or used as a vegetable.

DULSE *(Rhodymenia palmata)*

This seaweed, among the commonest of the red varieties, occurs all around the Atlantic, the Pacific coasts of America and Australia, and the Mediterranean. A dark-red or red-brown plant with a single deeply divided frond 8–40cm (3–16in) long and no stem, it is found on rocks and stones – sometimes attached to larger seaweeds – on middle and lower shores and in shallow water. Easily digestible, the dulse is best eaten raw, or can be added to stews or used as a vegetable. After a thorough rinsing in seawater, it may be dried for storage. Collect plenty, as it shrinks considerably when drying. Dried dulse can be fried or boiled.

LAVER *(Porphyra umbilicus)*

Found on the coasts of the Atlantic, Pacific and Mediterranean, the laver grows on rocks, large stones and timbers between the middle and low-tide zones, as well as on sand–covered rock surfaces. It has a thin, leaf-like, folded appearance and varies in colour between red, purple and brown, although it turns black as it dries. It can be eaten raw, boiled long and gently and/or fried.

IRISH MOSS *(Chondrus crispus)*

Also known as carragheen, Irish moss occurs on temperate coasts on both sides of Atlantic, where it grows on rocks on the lower shore and in shallow water below the tide line. Red-purple to purple-brown (or it may be green in bright light), its many branches are tough and leathery and grow to 3–16cm (1–6in) in length. Boil it before eating, or stew it with fish, meat or vegetables.

Sea lettuce

Laver

Irish moss

183

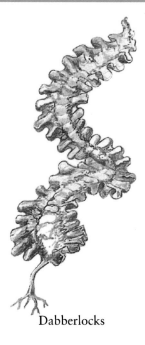

Dabberlocks

Sea kale

DABBERLOCKS *(Alaria esculenta)*
A North Atlantic seaweed, this olive-green to dark-brown variety grows on the lowest shore and in shallow water. A short stem secures a long, blade-shaped frond, with a flattened midrib, 10–90cm (4–35in) long and 5–15cm (2–6in) wide. This stem carries finger-like spore producers. The midrib and the stem can be eaten raw or in soups. The frond, with the midrib removed, should be soaked in fresh water for twenty-four hours, then boiled.

OARWEED *(Laminaria digitata)*
An inhabitant of temperate Atlantic and Baltic coasts, the oarweed is found in the zone from the low-tide line to water 4–5 metres (13–16 feet) deep. It grows on rocks and stones, often in large colonies. Olive to dark brown, it grows to 2 metres (6.5 feet) in length, with a frond divided into leathery, strap-like strips and set on a stalk 10–50cm (4–20in) long. It can be used in soups after preliminary boiling.

SUGAR KELP *(Laminaria saccharina)*
Found in the Atlantic and Pacific below the tide line on rocks and stones, this seaweed has the appearance of a crinkled sword blade and grows to 3 metres (10 feet) in length. The frond is chestnut to olive brown, and attached to a stem up to 40cm (16in) long. Smaller plants are best for eating. Young stems can be eaten raw, but the whole plant is edible if boiled.

The two plants described below grow near the seashore in the temperate regions, but are not seaweeds. However, they are of interest to the survivor in that both are edible.

SEA KALE *(Crambe maritima)*
Found on cliffs, rocks, shingle and dunes on north temperate coasts, sea kale is a perennial which grows to 60cm (24in) in length. Its lower leaves can reach 25cm (10in), while its upper leaves are much smaller; both are blue-green, smooth and have wavy margins. Small,

white, clustered flowers appear between February and May. The plant's shoots and young leaves are edible if immersed briefly in boiling water, then chopped and boiled in different water for twenty minutes.

SEA ARROW-GRASS *(Triglochin maritima)*
Found near north temperate coasts, salt-marshes and grassy areas near the shore, this plant has narrow, erect leaves 5–30cm (2–12in) long which grow from its base. The small, green flowers, which appear between July and September, form a loose spike on a stem 15–45cm (6–18in) tall. Sea arrow-grass can be eaten raw, but it is best added to soups or boiled as a vegetable.

Sea arrow-grass

FUNGI

Opinion about the use of fungi as a survival food is sharply divided. On the one hand, it is undeniable that many fungi provide food high in nutritional value, that only two to three per cent of all species are poisonous, and that fungi often occur in places where other edible plant foods are scarce.

On the other hand, the unwillingness of some to include fungi in the survival diet arises from these facts:
- Of the small percentage of poisonous species, some are deadly if even a small amount is eaten.
- Any symptoms of poisoning may not occur until

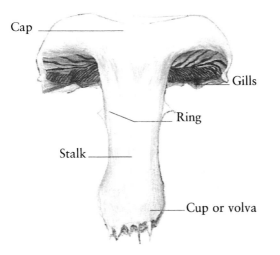

The largest group of edible mushrooms is the gill fungi, of which an example is seen here in stylized form. Not all species in this group have a ring on the stalk or a cup at the base.

> ### Survival tip
> About eight years ago I ate a wild mushroom of a kind that I had eaten many times before. Within twelve hours I was ill in hospital suffering from serious mushroom poisoning. As soon as I was able, I returned to the spot where I had gathered the fungi. They were still there and they were one of the edible varieties. So take care.

10–40 hours after eating, by which time only hospitalization offers any hope of recovery, should the poison be life-threatening. Even a relatively mild attack can lead to serious repercussions in a survival situation.

● This delay in the appearance of symptoms, combined with the extreme virulence of some species, means that there is no Edibility Test suitable for use with fungi.

● Individuals react differently to ingestion of fungi, even of non-poisonous species. The fact that one survivor eats a specimen with impunity does not guarantee that anyone else can do the same.

The only way to make safe use of fungi is to learn to recognize a few species so that they can be identified beyond doubt. The only way to identify any fungus is by visual examination and comparison with examples illustrated and described in a reliable guide. Even then, some authorities advise, beginners should have their initial identifications confirmed by another person with experience of edible fungi. This is good advice, since reference books themselves disagree, from time to time, as to whether a specimen is edible or poisonous.

Fungi, like mosses, ferns and seaweeds, are flowerless plants. They do not produce seeds but propagate by means of spores. Fungi do not contain chlorophyll, the green colouring matter in plants, and this is why they live on organic matter, either living or dead. The following catalogue contains some of the species known to be edible which are also fairly common and widespread. To make use of any of them, study their characteristics with care, and take every opportunity to identify them positively in the field.

EDIBLE FUNGI

HORSE MUSHROOM *(Agaricus (Psalliota) arvensis)*
Found in clusters or small rings, scattered though parks, gardens and pastures, the horse mushroom is very common between June and November. The cap is 5–15cm (2–6in) in diameter and egg-shaped with a flat

Horse mushroom

top. Bruising will cause the off-white cap to change to a yellowish-brown colour. The gills are a greyish-pink in colour, turning to chocolate-brown as the fungus matures. The stem is 10–13cm (4–5in) high and swollen at the base; the veil has two membranes. The flesh, which is firm and smells of almonds, has a nutty taste and is excellent to eat.

BOLETUS BADIUS

Similar in appearance to the cep, this mushroom favours sandy soil, and is mainly found in pine woods, where it grows in groups. The cap, 5–8cm (2–3in) across, has a slightly sticky and dark olive-brown top and a golden-yellowish underside. The stem is bulbous at the base and dark brown turning to yellow where it meets the cap. Under the cap the flesh varies from white to pale yellow, but stains to pale green if bruised. This mushroom is edible.

Boletus badius

FIELD MUSHROOM *(Agaricus (Psalliota) campestris)*

Commonly occurring in pastures and meadows between May and November, the field mushroom can be found individually or in 'fairy rings'. The cap, 5–10cm (2–4in) across, ranges from firm and round to flat-topped, and the gills vary from pink to light-brown, darkening with age. The stem is short and white. This mushroom gives off a fresh, pleasant smell, and is very good to eat. It is best picked early in the morning, when the dew is still on the ground.

Field mushroom

CEP *(Boletus edulis)*

Instead of gills, the cep has layers of pore-like tubes. It is found on leaf-mould in broad-leaved woods, and, between June and October, in large numbers. The cap, 10–15cm (4–6in) in diameter, looks like a burnt bread cob, giving the mushroom the nickname of 'penny bun'. The underside is covered in minute tubes from yellow to light green in colour, and the stem is 15–18cm (6–7in) high, with a light, vein-like cover. The white inner flesh has an attractive nutty taste.

Cep

HORN OF PLENTY *(Craterellus cornucopioides)*
This mushroom grows in large numbers in broad-leaved woods among the leaf-mould, arriving late in the year, from late August to November. Measuring 5–10cm (2–4in) in diameter, the dark-brown cap is funnel-shaped with wavy creases, and flows directly down to join the stem. Like the cap, the stem is ribbed and wrinkled. The spores are greyish. Although leathery and unpleasant-smelling, the flesh is very good to eat.

CLITOCYBE CONGLOBATA
Found growing in rich soil on country roadsides and in gardens and parks, this fungus has a cap 5–8cm (2–3in) across which is slightly convex and wavy. Grey to light brown in colour, it often has spots. The gills are slightly decurrent. The stem is short and curved and 2.5–8cm (1–3in) high, and unites at the base with others, to form a bunch typically consisting of 200–300 specimens. The flesh smells and tastes very pleasant.

BEEFSTEAK FUNGUS *(Fistulina hepatica)*
This fungus grows on old oak trees and occasionally on chestnut, and is fairly common from August to October. Growing singly or in pairs, it is tongue–shaped and grips the tree trunk at a height of 2–3 metres (6.5–10 feet). The upper crust is brown and tough, while the underside is normally lighter in colour and much softer. The inner flesh has a red, fibrous look, not unlike raw beef. It smells sweet and, despite what some books say, tastes good when fried.

PARASOL MUSHROOM *(Macrolepiota procera)*
Most often seen growing in fairy rings in pine woods and parks, the parasol mushroom can be found from August to October. The cap, 8–15cm (3–6in) across, is dry and smooth, and the thin stem is 15–18cm (6–7in) tall. Dry to the touch, the flesh is white, but bruising will cause it to turn reddish. It tastes of almonds and is excellent to eat, but not in great quantity.

Clitocybe conglobata

GIANT PUFFBALL *(Lycoperdon (Calvatia) gigantea)*
This fungus grows mainly in open grass areas, where it can be found between May and October, although it is not very common. More egg-shaped than round, it reaches 15–51cm (6–20in) across and sometimes even bigger, and can weigh as much as 9kg (20lb). The outer skin is white, smooth and leathery, while inside the flesh is soft and spongy, ranging from white to olive green at the centre. The flesh of young specimens is good to eat when fried. It will dry if left in the sun and will last for up to two weeks.

Giant puffball

CHANTERELLE *(Cantharellus cibarius)*
Growing in woodland, amid grass and moss, in large clusters, the chanterelle is found from June to November. The cap is 5–8cm (2–3in) across, convex and funnel-shaped, and the thick, vein-like gills run down to merge with the stem. The outer colour is a rich orange-yellow, while the inner flesh is white. The latter, which smells of apricot and has a nutty taste, is very good to eat.

WOOD BLEWIT *(Lepista nuda)*
This commonly occurring woodland fungus grows in rings, and is found between October and December. The cap, 8–13cm (3–5in) across, is smooth and turns from bluish-lilac to brown. The gills, tightly grouped, are pale lilac to brown, and the stem, which is slightly swollen at the base, is 8–10cm (3–4in) high. The flesh, which displays a slight violet-bluish tinge, has an aromatic smell and is good to eat, if rather bland in flavour.

Wood blewit

POISONOUS FUNGI

The majority of cases of fungus poisoning have been caused by eating species of the Amanita family, many of which are deadly, even in small quantities. The illustrations show some of the common features of this family, and any mushroom possessing any of these should be discarded. Look particularly for a volva, or cup, at the base of the stem, white gills, or scales (often pyramidal

Death cap

Destroying angel

in shape) on the cap. The typical smell of these fungi is like raw potato or radish. They grow singly in woodland on the ground, but never in fields or other open grassy spaces. Study the Amanita family with the help of a good guide, comparing them carefully with all the edible varieties you may wish to identify.

DEATH CAP *(Amanita phalloides)*
Found in deciduous woodland, parks and in humus-rich soil, the death cap appears in large numbers between August and September. When fully grown it has white flesh which gives off a sweetish and rather nauseous smell. The cap, 8–10cm (3–4in) in diameter, is sticky at first, but becomes smooth, with white patches of veil loosely attached. The gills are white to greenish in colour. The stem, 8–13cm (3–5in) long and 1cm (0.4in) thick at the base, is surrounded by a white, leathery, lobed volva. This species is deadly

DESTROYING ANGEL *(Amanita virosa)*
The destroying angel is found mainly in mixed or birch woodlands, between August and October. It does not normally grow in clumps, more often occurring in a scattered pattern. The cap, 5–10cm (2–4in) in diameter, is sticky and varies in form from egg-shaped to conical or bell-shaped. The gills are pure white and the stem has a cup-shaped volva at the base with a patchy edge. The white flesh, which has an unpleasant smell, tastes bland at first but very soon produces a burning sensation. This is a lethal fungus.

VERDIGRIS AGARIC *(Stropharia aeruginose)*
This fungus is easily recognizable by its pretty verdigris-green colour. Not very common, it is found in beech woods between July and November. The cap measures 2.5–5cm (1–2in) across and is very slimy. The gills range from violet to dark brown as the fungus matures, while the stem is a pale greenish white with a white ring. The crushed mushroom is greenish white and smells strongly of radishes. It is poisonous.

FLY AGARIC *(Amanita muscaria)*

An inhabitant of pine and birch woods, where it often grows in large clumps, the fly agaric is found between August and November. The scarlet cap measures 8–20cm (3–8in) across, and is sticky and covered with white, wart-like scales from the veil. The gills are white and minutely toothed at the edge, and the white stem is long and hollow, with a bulbous base. The flesh ranges from lemon-yellow to white. This species is poisonous, causes hallucinations, and can be fatal.

SICKENER *(Russula emetica)*

This very common fungus is found in most types of woodland and occasionally in scrubland, between August and November. Its cap, which often has a central depression, is 8–10cm (3–4in) in diameter, slimy, bright red and shiny. The gills are white and free from the stem, which stands 8–10cm (3–4in) high and is very brittle, splintering rather like bone. The sickener gives off very little smell, and is lethal.

DEVIL'S BOLETUS *(Boletus satanus)*

Occurring mainly in the chalky areas of southern England, the devil's boletus can be found between July and October. The cap, 13–15cm (5–6in) across, is pale to whitish green with bright-red pores. The stem is thick and bulbous at the base and covered with a coarse network of veins. Little odour is given off but the fungus can be easily recognized when crushed, because the white flesh turns blue. It is a poisonous species, causing violent symptoms.

PANTHER CAP *(Amanita pantherina)*

Commonly found in most deciduous woodland from August to October, this fungus is also known as the 'false blusher'. Its cap, measuring 8–10cm (3–4in) in diameter, is brownish with irregular white blotches. The gills are free from the stem, which is about 10cm (4in) high and swollen at the base. Where the stem emerges from a volva, it has a white, almost silky look.

Fly agaric

191

Red-staining inocybe

The crushed fungus is white and smells like raw potatoes. It is poisonous, and possibly deadly.

RED–STAINING INOCYBE *(Inocybe patouillardii)*
This is a particularly dangerous mushroom, causing almost certain death if eaten even in small quantities. Fortunately it is easily identifiable by simply touching it, when the pure, silky white flesh becomes stained with blotches of bright red. The cap is 2.5–8cm (1–3in) across, has a pink flush, and is fleshy. Its shape varies from conical to flat, but in all cases the edges are lobed, producing an irregular shape. The fibrous stem has no ring and is firm. The gills are at first white but later turn olive-brown with white edges. The crushed fungus gives off a fruity and not unappealing smell which can be misleading, since this is a deadly species.

SAFETY CHECKLIST

Since the risk of poisoning is ever-present with fungi, it is strongly recommended that, in addition to familiarizing yourself with these descriptions, you follow these guidelines:

● Do not collect old fungi, or any which are infested with insects or have been partly eaten by insects or maggots.

● Avoid very young specimens, especially if still in the 'button' stage, because many of the identifying characteristics do not develop until the fungus approaches maturity.

● Boil any fungus before eating it, and discard the water. There are some species whose poison is destroyed by boiling. Note, however, that there are others – for example, those of the Amanita family – which remain poisonous after boiling.

● Always examine any fungus carefully for the presence of a volva, or cup, at the stem's base before picking the specimen. There is a considerable risk that the volva could be destroyed, damaged or left in the ground when the fungus is picked.

● Ignore or discard any fungus with a volva at its base. Do the same if there is a ring of scales on the base of the stem, or if the top surface of the cap is speckled with small white patches or scales.

● Avoid any specimens which are red on the underside of the cap, or are producing reddish spores.

● Never eat a fungus with white gills, nor any gilled mushrooms with milky juice.

● Full recognition is the only safe way. Reject any other method of determining whether fungi are edible of poisonous.

11

MEAT FROM THE WILD

WEAPONS FOR HUNTING

AMBUSHES AND TRAPS

TRAPS FOR LARGER GAME

Mammals, birds, reptiles, fish, crustaceans and insects can all provide sustenance in a survival situation. Most of them supply the survivor with food of much higher nutritional value than do the great majority of edible plants. The most immediate problem is how to catch them. There are two basic solutions – hunting and trapping. If you have some experience of the former you may decide to adopt this approach.

While it is possible to kill animals with a club, spear, or bow and arrow, it does require skill. If you do not already possess at least rudimentary hunting know-how it is unlikely, even in a survival situation for an extended period, that you will become proficient enough in the techniques of hunting with improvised weapons to guarantee an adequate supply of prey. And in any case, unless you have weapons with you that are suitable for hunting you will need to improvise them. For these reasons you may choose instead to employ the equally ancient tactic of traps and snares to catch your food.

WEAPONS FOR HUNTING

If they have the right skills, making and using simple weapons will be a valid option for some survivors. This is particularly true in some very remote parts of the world where the wildlife is not normally disturbed by human presence and therefore have not yet learned to be afraid of man. It will thus be possible to approach any animals quite closely and to kill them without great difficulty. However, before looking at making the weapons that you will require, first we should consider the basic principles of hunting animals for food in a survival situation. The main problem confronting the survivor is to hunt successfully with primitive, improvised weapons. For this he needs to deploy skills of tracking and accuracy of fire which, if he had a firearm, would not prove so critical. Nevertheless, armed only with a few simple hunting weapons, he can improve his chances of success greatly by taking account of the following general rules :

- Early morning and late evening are the best hunting times, since animals will be moving between feeding and bedding grounds and their water source.
- All animals survive by being constantly alert. Their senses are more acute than ours. Also, their physical capabilities are generally superior, so that, for example, they can run faster and longer, jump higher, and climb trees or rocks more efficiently. The only chance a human hunter has – and this is true of many other aspects of survival – is to use his intelligence to offset his physical deficiencies. Observation, camouflage, a low profile, silence, smooth, careful movement downwind while the animals are feeding, and considerable patience – these are all indispensable if the hunt is to be productive.
- Any wounded animal is liable to be extremely dangerous, and the greatest caution must be exercised when returning to traps or snares.
- Never discard any part of a carcass without careful consideration. Nearly every part is of some use. Skins can be used as clothing, bones as arrowheads, needles or fish hooks, intestines provide gut, and sinews make thongs. In a survival situation you will

CONSERVING ENERGY

When seeking to obtain animal food from the wild it is vital to adopt an approach that is best described as 'cost-effective'. This simply means that the time and energy you put into hunting (and the same applies to all forms of trapping) any animal must be weighed against the value of the food it supplies. There is little profit in even a successful hunt if the hunting expends more energy than the resulting food puts back.

A long bow is most efficient if made from seasoned wood, which retains its springiness. Unseasoned wood can be dried slowly by a fire to improve its performance. Grease the bow with oil or fat if it is available, as this will also increase its flexibility.

Tie one end of the drawstring to the bow, then bend the shaft and loop the other end into a pre-cut notch. Release the tension when not in use.

find many more uses for what would in normal circumstances seem like waste.

The materials – natural or otherwise – that your environment can furnish will determine the type of weapons you can produce, but whatever the environment two straightforward, but important, rules are worth observing. First, keep your weapons simple but effective. Second, be sure to make full use of your imagination whenever you are confronted with a problem – the answer may not be a conventional one, even so it may prove perfectly efficient.

Simple weapons have been used for hunting since early humans first picked up stones and threw it at an animal, or brandished logs to fell prey. Indeed a small number tribes in undeveloped areas of the world still support their families by using home-made weapons for hunting.

LONG BOW

Provided that the materials are available, it is possible to construct a long bow very quickly. Take time to select the stave for the bow, making sure that it is good and strong, and using seasoned wood if possible. The best wood to use is yew, but any of the harder woods – for example, hickory, oak or birch – will do. Select a section that has no side shoots growing out from it.

Although customarily we call this weapon a 'long' bow, it is better if you construct one of no more than 130cm (50in) in length. Take your chosen section and taper off the two ends for about the last 50cm (20in). Traditional English bow-makers tapered a bow's ends so that they were round in section rather than flat, making both ends as even as possible to stop them twisting when the bow was drawn. This is the best principle to follow.

The next stage is to dry the bow slowly over a fire, a process which can take two or three days. Turn the bow over the fire as you would a roast pig. Under no

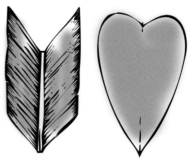

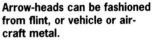

Arrow-heads can be fashioned from flint, or vehicle or aircraft metal.

circumstances should you allow the wood to burn. Once the bow is dry it can be greased using a generous amount of any kind of oil or animal fat.

To string the bow, first flex it to see if it gives naturally in any direction, and if so mark a small line on the inside centre of the flexed weapon. On the opposite side of your mark, at the tips, cut a small groove as illustrated on the opposite page, being careful not to cut too deep. Next use whatever string is available, as long as it is very strong. Parachute cord from a survival kit is excellent. The North American Indians used animal intestines which had been cleaned, dried and twisted together. Tie the string at only one end of the bow, and make a loop at the other end that can be slipped over the groove when the bow is flexed. Only flex the bow and put it under pressure when you intend to go hunting.

As for arrows, make ten or so to start with. Choose any strong, firm wood of about 65cm (26in) in length. Strip it of bark, then straighten it by gently bending it. Chewing it gently between the teeth is also a good way to do this. Carry out this process thoroughly because the straighter your arrow the better it will fly.

Next balance each arrow on your index finger so that there is an equal length either side of it. One end will be slightly heavier than the other and tip down. This will be the point end, while the lighter end will carry the flight. Carefully split the flight end to a depth of about 15cm (6cm) and slide in the flight, then carefully bind the split ends together with a light cotton, fishing line or fine snare wire. Flights can be made from any light material – for example, cardboard, plastic, polythene and, of course, feathers. Make the flight 10cm (4in) long and 5cm (2in) wide.

To make the arrowhead hard you will need to turn it slowly in a fire. Once this is done remove any charring and sharpen the tip.

At this stage you should practise rigorously with your bow and arrow. Once you have become proficient with the weapon, gently split the end of one good arrow and bind to it a scalpel blade from your medical kit. With an arrowhead of this kind I once managed to fell a fox.

CROSSBOW

Archery, using a bow like the one described above, is an age-old skill and very effective. However, if you have the time and the materials, it is worthwhile constructing a crossbow. Despite the rather unfavourable verdict that history has passed on it, the crossbow is lethal, and in a survival situation will greatly improve your chances of killing wild animals.

SLINGSHOT

A slingshot is very easy to construct from small amounts of material from your clothing. It is basically two 35cm (14in) lengths of cord or leather and a small pouch to hold your projectile. Form a loop big enough to fit over your finger in one of the cord ends and a knot in the other. Slip the loop over your index finger and hold the knot between your index finger and thumb. Place a smooth 3cm (1in) pebble in the centre of the pouch, making sure it is secure. Gently swing the sling in an arc at your side then bring it up more quickly above your head.

Concentrate on the target, and at the moment of release let go of the knot and throw your arm in a straight line towards your target. To begin with, it can be quite difficult to achieve accuracy with the slingshot, but if you are prepared to practise it can prove to be a very effective weapon.

Survival tip
Unless you have good reason to do so, do not lash a knife to a spear. On a survival exercise in France I cornered a sheep, at which I hurled a spear tipped with my knife. I hit the target, but unfortunately the sheep was not killed, and ran off at high speed, taking my only knife with him.

Concentrate on the target. not the slingshot, and do not swing too wildly before you release it.

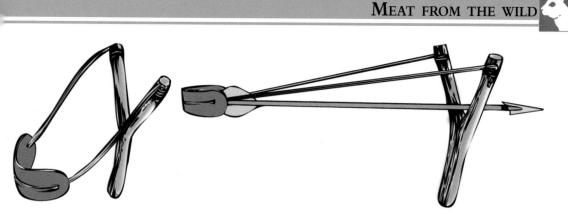

CATAPULT

Many of us learned how to use a catapult in our childhood. The bonus for the survivor is that it is very easy to construct one. The only man-made item required is rubber or elastic, and for this reason I suggest that a good length is worth including in your survival kit. However, if you find yourself with none but a vehicle is at hand, the rubber from an inner tube is excellent. But for a catapult that will be of any use, clothing elastic is by no means strong enough.

The principle of the catapult is so straightforward that I need not go into the details of its construction. However, I must stress the importance of securing the elastic properly.

If you have enough elastic, try using an arrow instead of a stone. This will give you a weapon with much greater accuracy and killing power.

GLASS-KNIFE

To fashion a simple knife, wrap cloth to form a handle around a 15cm (6in) shard of glass.

SPEAR

Unless you can corner your prey, a spear is of little use in hunting. Nor is at all easy to throw one with much accuracy and force. However, this weapon will serve the very useful purpose of offering protection against attack by a wild animal – either prey or one that acts aggressively without provocation – and for this reason it is a good idea to make one.

A catapult is an easily made and worthwhile addition to your hunting armoury, although to become an accurate shot takes practice. Include a length of suitable rubber in your survival kit.

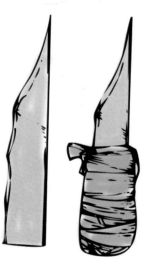

A knife is essential in survival: Make one by heating a piece of metal in a fire and beating it into shapewith a stone. Alternatively, a shard of glass can also make a simple but effective blade.

A spear-thrower (below) will add force to your weapon. Split a section of a branch of about 5cm (2in) in diameter, which has a small offshoot to be used as a handle, and gouge out a smooth channel for the spear. Leave the end blocked to act as a buffer.

Construction is simple — a reliable spear can be as basic as a good strong staff some 180cm (70in) long with a sharpened end. However, if suitable material is available, a spearhead can be formed to improve efficiency. If metal from a vehicle or aircraft is to hand, it can be beaten into the required shape. Alternatively, use a sharp piece of stone – flint is particularly good. Whatever the material used for the head, it must be bound securely to the spear, to ensure maximum impact on hitting the prey.

STONE AXE

For the axe handle select a piece of hardwood of about 60cm (24in) in length. Split one end centrally to a depth of about 20cm (8in). Using one hard stone to chip away at the other, sharpen an axe head from a stone such as flint, quartz or any other hard, glassy rock. Insert the head into the split in the handle so that it fits tightly. The tighter the fit the more effective the tool. Be very sure to protect your eyes when you are chipping the axe head into shape. With each strike tiny fragments will be sprayed about and these can easily blind you.

Spearheads can be made in a similar way to arrowheads. It is vital that the head is fitted securely to the staff.

SHARPENING A KNIFE

In most cases tools are required to make weapons, and for the survivor this will mean using his knife, or items from his escape kit, or more likely both. Therefore it is important to look after all your equipment and in particular to sharpen your knife daily. To sharpen a knife, find two pieces of sandstone and rub the two flattest surfaces together to form a sanding block from one of the pieces. Move the blade in a clockwise motion over the block, with the cutting edge moving away from you, and applying slight pressure. Ensure that the movement is constant, and not at all jerky. Wet the surface of the block as required. Be sure to give equal attention to both sides of the blade.

AMBUSHES AND TRAPS

For those with no experience of hunting, or little inclination to rely on it in part or at all, the only viable alternatives – apart from subsistence on plants (see previous chapter) – are simply ambushing animals or capturing them in a trap of some kind. Indeed some survival experts would say that unless you are one of the few people who possess the skills to stalk wild animals and get near enough to them to dispatch them cleanly with simple weapons, your best chance of success lies in taking your prey in one of these ways. (You will still have to kill your animal, of course, but you will have bypassed the more difficult aspects of hunting with a weapon.)

Look for animal tracks, or 'sign', leading from feeding grounds to water. Early morning and late evening are the times when most animal movement occurs. Smear mud on your face and hands to break up their outline and help them merge into the background.

Survival tip
On an exercise in Canada, I found it possible to walk quietly up to wild partridge and pick them from their nesting place in the trees.

Most animals are creatures of habit. Regular observation will enable you to predict the behaviour of each species, and to plan your hunting or trapping accordingly.

Conceal yourself thoroughly some time before you expect, from your previous observations, any movement to take place.

The creatures, mainly mammals, that you are most likely to seek in a survival situation are discussed below. Little description is given of the mammals included here because for the purposes of taking them the survivor needs to know little more than that they range from small to very large and leave tracks that differ widely. It requires little knowledge to determine which species of mammal has left a particular track. One very easy conclusion to draw is that large paw impressions must indicate a creature that is both sizeable and heavy. In such cases, try standing close to the sign and compare the depth of your own footprint. If the animal's track is bigger and deeper than yours, it is best to leave it alone.

Deer, goats and sheep Although the tracks vary in size, the basic shape of the sign left by deer, goats and sheep is more or less the same. Most of these creatures are fairly safe to hunt, but you will require time and skill. Approach downwind, and remember that camouflage is more important than sound.

Wild cats and dogs The sign left by wild cats and dogs is fairly easy to identify. However, both types of animal can be very dangerous, and the latter particularly so because they normally hunt in ferocious packs. If you intend hunting them, familiarize yourself with the techniques used in dog evasion.

Whenever you are out hunting animals or foraging for edible plants, watch out for evidence of small game – for example, rabbits, birds and waterfowl, squirrels, rats and mice, frogs, snakes, insects, snails and worms. Many of these will reveal their presence by their nests, burrows or dens, as well as by their tracks and runs, meal remains, droppings, territorial markings and possibly calls, songs or even smells.

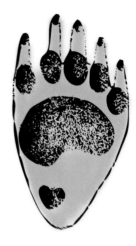

Bear prints (usually 15-30cm across) show foot- and toe-pads as well as claws.
Cat prints are smaller (4-10cm across) and more rounded, showing pad and toe.

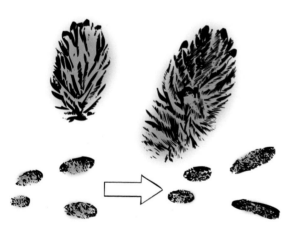

Rabbits From the Arctic circle to jungles and deserts, rabbits can be found in most parts of the world. They are fairly simple to trap and make delicious eating. However, there is very little fat on these creatures and for this reason other animals should be trapped to supplement your diet.

Rabbits can generally be found in large numbers, giving a good food supply. Their print, although light, is very distinctive.

Covering fresh eggs with fat will preserve them for up to two months.

BEES

Along with eggs, honey is the oldest food known to man that is still consumed in its natural state. It is highly nutritious and has many uses in food. Once the honey has been drained from the comb, the wax can be used for making candles or water-proofing clothing. You will need to build a smoke fire under your nest to collect the honey. Bear in mind also that pupae, larvae and adult bees are all edible.

Birds All birds are edible, but some species, especially those found near coastlines, are so disgusting in taste they are difficult to digest. Inland, larger game birds are quite easy to catch on the ground and all make good eating. Birds' eggs are a safe, nutritious food, so don't neglect them. If you observe birds nesting, check the nest for eggs. All eggs are edible and provide a good balanced meal when eaten with, for example, stewed nettles. However, beware of angry and protective birds such as seagulls – they are very likely to attack you.

Rats Despite their bad reputation, rats make good eating, and in parts of South America, where they are a major problem, they have been processed and canned for consumption. And, contrary to what is commonly claimed, they do not eat rotten food. However, they do carry germs and lice. Thus once you have trapped and, killed the rat, skin it and then discard the rump end, intestines and head along with the skin. Unsurprisingly, rats are common all over the world.

Frogs Most frogs are edible, but note that most of the meat is on the legs. They are easily caught, by tapping the water gently with a flat piece of wood; when the frog appears, tap it on the head. Frogs will reveal their whereabouts at night by croaking. Patience is needed at the outset, for if they are plentiful, they can always be heard somewhere other than the place you are.

Snakes Most of the flesh of snakes is edible, and it is very good to eat. The problem lies in finding and catching your prey. Snakes are experts at camouflage and will in any case shy away from humans. On sunny days they can be caught out in the open and should be pinned to the ground with a forked stick. Remove the head, strip off the skin, split the snake down the middle and remove the centre gut. Roast or fry the meat.

Snails Some snails are dangerous to eat, but fortunately most of these species are identifiable by their

Always kill the rat before picking it up; remove both fur gut before cooking.

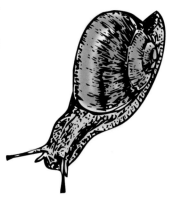

bright colouring. (If in doubt about its edibility, avoid a snail.) Snails found in Britain are all good to eat. They are rich in nutrients, yet it is inadvisable to eat more than a dozen or so at a sitting. Starve your snails for a few days or leave them in a bucket of salt water. This will purge their system. Boil them for about twenty minutes, but don't overcook them otherwise they will go hard.

Worms The thought of consuming worms puts many people off, but the fact remains that they make good eating. The simplest method of preparation is to dry them in a billycan by the fire, shaking the can from time to time to stop them burning. Once dry they can be crushed to a powder which provides a protein base for soups.

Snails and worms have excellent food value, and are usually very easy to find.

CATCHING SMALL GAME

Ambushing is a useful way of taking small game, but a more reliable and less energy-consuming method is to catch them with snares and other forms of trap. Properly made, and effectively sited and set, they will do the waiting – 24 hours a day. Any food they provide will be highly cost-effective in energy terms. To begin with, set out to catch small game, as they are among the easiest to trap, relatively common, simple to carry back to camp, and easy to prepare.

Rabbits are a good example. Their habitats are grasslands, open woodlands and dry, sandy localities in lowland places. They are social animals, always living in groups and usually remaining in one area. They occupy burrows, sometimes with a number of entrances. Banks and slopes with light tree or shrub cover are chosen for burrow sites.

Rabbits use specific routes or 'runs' between burrows and these are conspicuous. Runs, as well as the entrances to the burrows, reveal the animals' small, dark, round droppings if they are in cur-

TIPS ON COOKING MEAT

Animals larger than the domestic cat should be boiled before roasting or baking. Larger animals will need to be cut into manageable pieces before boiling. Roasting, unlike boiling, should be done as quickly as possible, since a slow fire's direct heat will toughen the meat. Meat that is very tough is best stewed with vegetables.

rent use. The most effective trapping method is to set snares (see below) in the runs a little distance away from the burrow entrances. This is because animals are naturally on the alert as they emerge from underground and are therefore likely to spot and avoid the snare. When setting snares, you must take care not to step on or disturb the run in any way, and rub the snare and your hands with animal droppings to help conceal your scent.

DRAG SNARE

The simplest snare, and the quickest to set, is the drag snare. A simple noose, it kills the animal by strangulation. The most efficient material is brass snare wire, but steel wire, nylon cord, strips of hide or even a wire saw are good alternatives. Whatever the material selected, use a length of about 50cm (20in). Make a loop 1cm (0.4in) in diameter at one end, and pass the other end through this. The snare is now ready. Check that the material slips freely, so that the noose tightens and opens easily, and if wire is used, that there are no kinks, which could catch and allow the prey to escape.

The snare should be securely fastened to a stake driven into the ground, or hung from any natural deadfall above the run. Set a noose about the size of a fist so that its lower edge is about 10cm (4in) off the ground. If the site is suitable, place dead twigs and branches on either side of the snare to help channel the rabbit into it. Do not use green twigs, as the rabbit may stop to eat them. Use the twigs to hold the snare off the ground if you are using anything other than wire.

When setting a trap:
– Avoid disturbing the ground;
– Set the snare covering the natural path of the animal, at about 10cm above the ground (for most small game);
– Do not make the snare so big as to allow the animal to pass straight through it.

It is a good idea to set a number of snares, but you should site them some distance from each other. When an animal is caught the disturbance it creates will instantly put all nearby animals on the alert, but its cries will not necessarily be heard by those animals likely to investigate distant snares. Check all the snares the following day, for it is cruel to allow any animal caught but not killed to suffer unnecessarily. In addition, if you leave it for too long, some other four-legged hunter may make off with your catch. Collect any animals caught and reset the snares.

There is little point in catching an animal only for a larger one to devour it. Construct a trap that lifts your catch above the ground.

BALANCED POLE-SNARE

Some useful modifications can be made to the simple drag snare. For example, it can be developed into a balanced-pole snare. This has the advantage that it lifts the snared game clear of the ground, ensuring that your catch does not become some other creature's meal.

To make a balanced-pole snare, tie a suitable pole to a convenient tree so that its lighter end can be pivoted downwards to be directly above the animal run. Then tie the loose end of a drag snare firmly to this end of the pole after knotting it, at a convenient point, around the movable half of a snare trigger. This can be made

A balanced pole snare is ideal for catching smaller game such as rabbits. Once the snare has been triggered, the rock weight should be sufficient to raise the trapped animal well out of the reach of other predators.

of two interlocking pieces of wood, which come apart when the trapped animal pulls on the snare. Fasten a weight (a rock is ideal) to the other end, so that when released, the pole holds the snare at least 1 metre (40in) off the ground.

Drive the other half of the trigger into the ground alongside the run. Engage the trigger halves and the

It is not always easy to snare your prey. Blocking the natural route is one way to try persuading the animal into the trap. This is best done where the animal travels to a watering spot close to its lair. Use old sticks to block the path as the animal may stop to eat fresh foliage.

snare is ready. Check that the loop runs freely and that the noose is set at the correct height above the ground for the intended prey.

HALF REEF KNOT-TRIGGER

If snares are being made from cord or nylon fishing line, a simpler variety of trigger can be used with the balanced-pole set-up. Fasten about 1 metre (40in) of cord to the light end of the pole so that about 10cm (4in) hangs down at one side and the remainder at the other. Tie a simple knot on the end of the short line. Use an overhand or bowline knot to form a 1cm (0.4in) loop at the longer end. Bring the light end of the pole down, and tie both ends of the snare line using a half reef knot around the anchor stake. Support this knot against the stake while allowing the pole to rise gently. The half reef knot will slip at first, but eventually the knot at the short end will bear against it and hold the pole in position.

Use the looped end of the cord or line to form the noose and set the snare. When an animal jerks the noose end, the half-reef knot will twist, release the other knot and allow the pole to rise.

SPRING SNARE

An alternative to the balanced pole is the spring snare. A small sapling or bent branch held under tension performs the function of the balanced pole. However, bear in mind, the half reef knot trigger is probably unsuitable for use with spring snares.

When the trigger is tripped the animal is lifted clear of the ground and strangles itself. If cord or nylon line is being used, the snare noose will have to be supported in some way to keep it open.

Spring snares can be constructed easily on site by doubling a sapling over and securing it with a simple snare and a trigger device. A portable bow-snare can be made from two saplings which are forced into the ground and bent over to form a bow. They are locked together by a bait pole. When this is moved in any direction, the two saplings – and the snares on them – spring apart.

TRIGGER VARIATION

Other types of trigger than those described above can be employed if longer lines are available. One of the simplest is shown below, in use with a spring snare. If line or wire is available, this trigger can carry a snare at each end.

Always keep your head turned to one side when you are setting any spring snare.

SNARE HOLE

Disturbing the earth by digging a hole will frighten most animals away, but if the hole is baited with meat some animals — for example, foxes, badgers and pigs — will soon come sniffing around. If you add spikes to the hole in such a way as to make the animal shake its head and wriggle to avoid them, the snare will be operated and tighten.

SQUIRREL SNARE

If you see any indications of squirrels, it is a good idea to set snares, but for these animals you will need snare wire rather than any other material. Signs of squirrels include pine cones stripped of their scales (except, perhaps, for a tuft at the top) and the scales themselves, scattered at the base of a tree. In deciduous woods, look for nut shells split in two and claw marks on tree

Squirrels climb with their heads down, so place several snares loops so that they rest on the pole.

trunks. Old mushrooms, wedged into forks near the trunk, also indicate squirrel occupation.

The first step is to provide the animals with an easier

route up into the tree. Do this by propping a suitable pole, 4-5m (13-16 feet) long, at an angle against the trunk, wedging it into a fork if possible. Prepare three or four snares, using about 60cm (24in) of snare wire for each, and setting nooses for all of these measuring about 7cm (3in) across.

Tie the snares on to the pole, using the extra length to form a curve so that each noose hangs vertically about 10cm (4in) above the pole. Leave 1 metre (40in) of the pole clear at each end. This is one set-up in which snaring one animal will not necessarily lessen the chance of catching others. These latecomers may well be attracted to the disturbance, decide to investigate it, and snare themselves.

BIRD SNARE

Small drag snares can be adapted to catch birds. Shown here is a method of hanging snares on and around a well-used perch. The signs of such a roosting sit are quite obvious: there will be large amounts of droppings on the branch or ground below.

Birds can also be caught using a snare on a baited perch. The square-ended perch must form a loose fit into the upright, being held lightly in place by the knot in the snare line. When a bird dislodges the perch the

Make a baited hole snare by digging a small hole in the ground and placing your bait in the bottom. Drive in several sharp pegs through the sides, so that they protrude on the inside of the hole, over the bait. Rest a secured snare over the tips of the sharp pegs.

A series of snares can be set in a tree on a branch where there are signs of birds habitually roosting.

To snare wild fowl and sea-birds half fill a bottle and secure a few snares and a little foliage to the neck,. Float the trap out on a secure line and check every morning.

wcight falls, as it is no longer held by the knot, and draws the noose around the bird's legs. Do not use too heavy a weight — 0.5kg (1lb) is sufficient.

BAITED HOOK

If you are near water it is always worthwhile to bait a fish hook and cast the line out a little way, tying the other end to something stable, such as a rock or a tree on the shore or bank. There is a good possibility that a bird will peck at it and hook itself.

Most birds will feed or nest in the same place daily.

Waterfowl may also be caught by a floating trap of the kind illustrated. If no vessel is available, a weighted piece of wood can be used instead.

FIGURE-FOUR TRIGGER

A little more difficult to make than the triggers already discussed, and perhaps less easy to set, the figure-four trigger compensates by being very sensitive, especially when used with the deadfall traps described below. Disturbance of the trigger causes the rock to fall — hence the term 'deadfall' — trapping the animal. One advantage of this trap is that the pieces of the trigger can be carried and used repeatedly if travelling.

It is always worth making several different-sized triggers, for catching different types of game. Once your 'figure four' is set, balance a large rock or log over the bait end. Check on a regular basis.

BAITED BOX

In addition to the snares and traps already described, some traps are designed to capture the quarry without killing it. The least complicated of these is the baited box. All that is required is the materials to construct a box or cage and something to bait it with. When an animal enters the trap and investigates the bait, it nudges against and releases a peg. This in turn allows the door to fall at the entrance of the trap.

A baited box is really a more elaborate version of the baited snare hole. It can be constructed in the ground or fashioned out of reusable materials. It is best to cover your box trap with foliage once it is constructed, but make sure that the bait can be seen from the entrance.

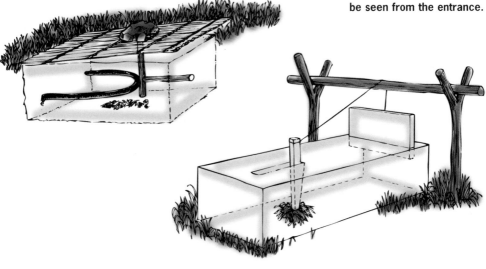

PURSE NET

If you have a purse net – or can make a gill net as shown in the next chapter – you will have another reliable method of catching small game such as rabbits. Having located a burrow, stake the net over an entrance that shows signs of recent use. Make sure that you have located all of the other entrances and block all but one of them up. Then blow smoke or pour water into the one remaining open hole. There is a good chance that the rabbit or other occupant will make a panic exit and bolt straight into the net. Even if

Pour water down another entrance of a rabbit burrow or similar lair. The panicked occupant will bolt – straight into a waiting purse net. Burrowing animals can be easily caught by blocking its escape route from the burrow with a purse net.

you do not have a net, you can use the same method but with a snare set outside the hole.

TRAPS FOR LARGER GAME

Large traps can be dangerous to the survivor as well as to the intended prey, and therefore great care needs to be exercised when using them. Because a considerable amount of time and effort is required to construct traps for such animals, it is wise to wait until you have sighted sufficiently large numbers of them to justify your investment.

DEAD FALL TRAPS

The trigger release line is used here with a trip-line operation, and baited if required. A variation is to use a snare attached to the tension line via the trigger, so that the deadfall trap becomes a bigger-game version of the balanced-pole snare.

The adaptable and inventive survivor might even combine the two operations — using one trap with two trigger lines, covering baited trip-line and snare operation.

DEAD FALL SPEAR TRAP

A development of the normal deadfall trap, the deadfall spear trap has the added refinement of a number of sharp spears. Also required is positive knowledge of

Large animals might be caught in your trap, but may also be strong enough to destroy it. Traps appropriate for such animals need time to construct and must have more accurate mechanisms, with a quick and simple trigger. Provided you are not too distant, you will be able to hear larger animals when they have been caught.

Take care when leaving dead fall traps. Make sure you are not beneath the trap when you are setting the trigger. If there are several people in your party, always ensure that they know where the larger traps are located.

UTENSILS

When you have caught your prey, you will require utensils in which to cook it, and with which to eat it. As with weapons for hunting and traps and snares, you will probably have to make these. The most important of all such survival items is a billycan, and no SAS operation goes ahead without each member being equipped with one. A billycan will not only allow you to collect water and edible plants, but will also enable you to cook the latter, as well as cooking meat, fish, vegetables and other foodstuffs, or simply to boil water.

If you originally set out equipped for the possibility of finding yourself in a survival situation, you will probably still have a purpose-made billycan with you. And if are in such a situation as the result of a transport failure or crash, you will most likely be able to retrieve either a billycan or another suitable receptacle from the vehicle or aeroplane in which you were travelling. However, if neither possibility is open to you, there is still a good chance — unless you are in a very unfrequented environment — that you will be able to find a discarded tin can of some kind. The larger this is the better, although even a soft-drink tin can be used to boil water over a fire, and thus to cook. As for utensils for eating and drinking, a simple spoon can be carved from a flat piece of wood with a penknife, while a mug can be fashioned from a section of bamboo, where this is available. If neither wood nor bamboo can be found, use whatever material is to hand.

Where bamboo is present, one other item that will prove useful is a bamboo shovel. Use this to dig your hangi fire pit, latrines, or spear pit, and even to make a hide. In tropical prison camps escape tunnels have been dug with them.

the animal's habits and considerable skill in concealing the trap. Note also that the better prepared this form of trap is, the more dangerous it becomes to the trapper himself — and even more so to fellow-survivors. Once triggered it stabs as well as clubs the quarry.

SPRING SPEAR TRAP

The trigger for the spring spear trap is bait hanging from a branch. The animal will need to jump for the bait or at least raise its head, and in pulling the bait it triggers the spring trap. The short spears are positioned so that they pass in line with the bait. The branch arm should be as springy as possible and under effective pressure, and the release trigger should be kept simple.

12

FOOD FROM THE WATER

ANGLING

Both fresh water and salt water provide a variety of foods for the survivor, the most important of which are fish. Nutritionally, fish are the most desirable of the aquatic foods, which also include crustaceans and seaweed. However, they are also the most difficult to catch.

The methods used to take fish – hook and line, nets, traps, snares and spears, even the hand – have all been in use for many centuries. Each technique calls for a combination of practical skill and some understanding of fish and their habitat. In short, you need to know not only how to use the equipment, but also when and where to use it. In addition, like every fisherman, you will need plenty of patience.

ANGLING

Strictly speaking, angling means using a hook and line attached to a rod, but in a survival situation where you have no purpose-made fishing tackle a simple rod can be made from a straight branch. In fact, even this item of equipment can be dispensed with, since you can simply hurl the baited hook into the water and hold or secure the other end of the line.

FEEDING HABITS OF FISH

Whether they live in fresh water or salt water, fish vary as widely in their diet and habits as do land animals. In addition, the time of day, the nature of the water and the weather all have an effect on their feeding activity. However, some generalizations about their feeding habits are widely accepted, and these concern the times when feeding tends to occur. The most important of these rules-of-thumb is that dawn and dusk are the two periods when fish feed most readily and are therefore the times when bait fishing will be successful. Even so, some species feed more eagerly at other times of either

day or night. Other favourable times for fishing are:

- When fish (in fresh water, especially minnows and other small fry) are seen rising to the surface or jumping clear of the water.
- When a river or stream is in full spate after rain. Most fresh-water species feed at this time, even though the water is muddy, in order to obtain any extra food which is carried down by the high water. A small backwater served by a minor tributary is a good fishing location at such times.
- When waterfowl are in a group on the surface, or are diving.
- When a storm is impending.
- Just before and just after a full moon.
- When fishing from the seashore, during the hours on either side of high tide.

It is also well known that fish react to hot weather by moving to cooler water where this is available – in effect, water which is deeper or shadier, or both. In hot weather on rivers, go after fish which are lying in the deeper pools and under shady banks. A combination of conditions conducive to feeding often occurs along the outer bank of a river bend, particularly if the river level is low. If you are using a natural bait, try to move it downstream at the water's pace, since in this way it will seem more lifelike to the fish. However, at the same time try to get it to pass near underwater rocks, logs, or undercut banks, because these are the places where fish seek cover from both hot weather and, in some cases, from predatory species. When you are fishing a still water in hot weather it is best to present your bait deep in the water, as the fish will almost certainly have descended to a cooler level.

In cooler weather – and at dawn or dusk – fish the shallower areas of rivers or streams, or around the edges of lakes or ponds. When there is less warmth and light levels are lower, the fish will search for warmer water – and in any case they tend to feed more readily

To encourage fish to feed in a specific area, suspend offal or carrion over the water. Ensure that it is well away from the bank so that scavenging animals cannot reach it. Blowflies will deposit eggs on the meat and these will turn into maggots and eventually drop into the water, attracting fish. The process will take several days, but in the meantime try fishing in the same spot with a baited hook. When maggots have developed, these too can be used as a hook bait.

in the shallower areas. In a stream or river, always let your bait move with the flow. Fish habitually lie in the water facing into the current, for two reasons. First, they can spot anything drifting downstream; second, this position ensures an easier flow of water over their gills, through which they breathe.

Allowing a natural bait of the kind described below to move in this way gives you the best chance of its being seen, accepted and taken. Some species, such as carp, tench, catfish and eels, feed on the muddy beds of lakes and slow-moving rivers. Bait should be cast on the bottom and moved very slowly to explore the water for these species, whose size provides more food than the smaller fish.

Use live baits native to the chosen water or its surroundings. Maggots, worms, grasshoppers and small fish are usually effective.

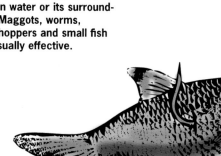

Lures simulate both the look and the movement of prey fish. Make them with at least one joint in order to emphasize the latter characteristic.

Make sure the hook passes through a fleshy part of the bait, but do not kill it. The bait's swaying action in the water will attract hungry fish.

NATURAL BAITS

Look for signs of insects, grubs, worms, minnows and other small fish, shrimps, shellfish and other creatures found in or near water. Make these your first choice for bait, since they will be the food to which fish are accustomed, and will not make them suspicious. Certain berries – elderberries, for example – will also take fish. If no natural bait can be found, make use of alternatives such as scraps of meat.

LURES

A lure is an artificial bait used to deceive predatory fish by simulating the appearance of a small fish or other food creature. You must contribute to this deception by pulling the lure through the water to mimic as closely as possible the movements of the creature it is intended to imitate visually.

In the earlier coverage of the survival kit it was recommended that you include a purpose-made lure in your fishing tackle. If, in a survival situation, you find yourself with one, all the better. If not, lures can be made from a tuft of hair (from an animal or your own), a piece of fin (from any fish) with a scrap of flesh

attached, brightly coloured scraps of cloth or pieces of feathers made up to suggest an edible creature. Each of these should be made up around the hook in such a way as to conceal it and look appealing.

A hook can be made from a safety-pin, a nail or even a large thorn. Add a button or coin to make a spinner. The brightness and revolving action will prove irresistible to predatory fish, which are in many cases larger species.

HOOKS

Your survival kit should provide you with sufficient hooks to fish for many months. Remember that big hooks will only catch big fish. This is because little fish can only nibble at, rather than take in, a bait big enough to mask a large hook. You will do better seeking a large catch of small fish rather than going after only their big brothers, which have grown big because they are wary and therefore harder to tempt. If

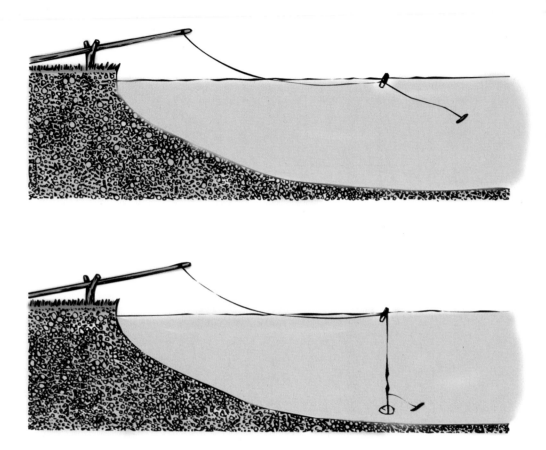

Fish feed at different depths in the water, and you may need to experiment to locate them. To present a bait near the surface, use a float without weights below it (top). The bait will sink to just below the surface under its own weight. To fish on or near the bottom of the water, add enough weights to sink the bait deep (above). The float should be mainly below the water, with its tip visible to signal a bite.

you have no manufactured hooks, you can make your own. Safety-pins, paper clips, split pins and stiff wire will all serve the purpose, but there are many other possibilities. All that matters is that the hook should be sharp enough to penetrate the fish's lip or the inside of its mouth when it takes the bait.

FLOATS AND WEIGHTS

Any small floating object, such as a cork or small piece of dry wood, will stop the line from sinking deep, which is what is required if the fish you are trying to catch are near the surface. A float also lets you know when a fish is showing interest in your bait. By adding weights to the line – if you have no ready-made ones, use a small stone – you will cause the line and the bait to sink. This will make it easier to fish the bottom of a river or still water, which is where fish move to when it is hot and where some species feed at all times.

GILL NET

If sufficient line is available – it might be provided by the rigging lines of a parachute, for example – you can make what is probably the easiest and most effective method of catching fish – the gill net.

When you have decided on the width of your net – 2 metres (6ft) is right for small streams – tie a line between conveniently spaced saplings or stakes to mark the top of the net, and one below to mark the bottom.

Suspended in a narrow stream, a gill net will trap fish by the gills as they try to swim through it. It is possible to make one from parachute rigging lines or other suitable line. This task is time-consuming, but the result will be a very effective way to catch fish – and most likely more than one at a time.

If nylon lines are available strip out their inner cores. Double a piece of this inner core and tie it to the upper line. Use an overhand loop, letting both ends hang down loose. These hanging strands should be about 30 per cent longer than the net depth required.

BASKET TRAP

There are several types of fish traps that a survivor can employ, and they may be used in either fresh water or the sea. Some make use of tidal movement, while some are best used in more restricted waterways. They can be permanently sited or movable, and some contain bait, while others do not.

A basket-type trap can be constructed fairly easily from light, flexible twigs, woven together. The cone-shaped entrance makes it easy for the fish to enter, but difficult or impossible to get out. The key to this pattern of trap is the 'lobster-pot' entrance, but the overall shape is immaterial. Placed in a stream, river or rock pool, a few of these traps will provide a good supply of fish. The effectiveness of this kind of trap will be increased if a suitable bait is tied inside.

It is possible to make the body of the trap from a piece of stout bamboo of about 1 metre (40in) in length. Wrap a few turns of binding around the stem a few centimetres from one end. Then split the full length of the bamboo from the other end as many times as is convenient. Forced apart, these split pieces will form a cone. The ends can be tied to a hoop with line, wire or vegetable fibre. Finally, add a funnelled top. This can

Design a basket trap to allow fish to enter but not to exit (top). A baited basket based on a lobster pot will, when left on the river bed, attract crayfish and eels – both edible.

Make an on-site fish trap for use in a river by driving wooden stakes into the bed. Damming the river will help to guide fish into the trap.

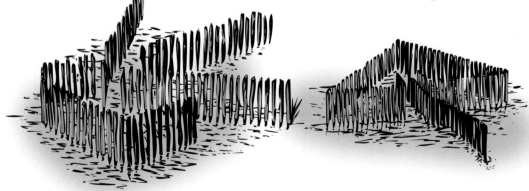

be made from a shorter piece of bamboo in the same way as the cone was constructed.

ON-SITE TRAP

Make the sides of the trap from wooden stakes driven into the river bed and, if necessary, tied together. An arrangement resembling split fencing and supported by additional stakes will be effective. The shape of the trap is not critical, as long as it incorporates a version of the funnel entrance. This kind of trap demands time and effort, but it will provide fish and is best used in conjunction with spearing.

SPEARING FISH

In daylight spearing fish with a sharpened wooden pole is by no means the easiest way to take them, but it can be done. It is essential that the spear is rigid and that the point is kept sharp. A blunt point will merely glance off the scales of any size-able fish. A three-pronged spear is more difficult to make, but it will prove far more efficient than a single-pronged version. Do not throw the spear. A stab or thrust is better controlled and more effective. Try to find a position over the fish and strike directly down-wards. This method has the twin advantage of mini-mizing the effects of refraction and of pinning any speared fish to the bed of the stream.

FISH SNARE

A fish snare is simply a wire loop on the end of a stick. A stick about 200cm (80in) long is ideal, although you may prefer to vary its length to suit your needs. The snare can be one of the snares recommended for animals (*see* page 205) with an extension to the cord

In clear water, and if the fish are big enough, try spear fishing. Combining damming with the use of a gill net and fish traps should ensure that there are fish to spear.

Catch large fish with a snare, either suspended or on a pole. The pole can be used for slower birds such as quail.

which will allow you to close the snare. Snaring fish is not easy, but, if you are patient and remain alert, it can be done. Observation is also important, and you should study the stream or river to identify the places where fish tend to lie.

TICKLING TROUT

Brown trout and rainbow trout are found in rivers and in stream-fed lakes and reservoirs used for fishing, but

Trout can be found in running water and in some stream-fed lakes and reservoirs. Often they lie close to the bank, where it is possible to catch them by the age-old technique of 'tickling'. Insert your hand into the water downstream of the fish – that is, behind it. Gently and slowly pass your hand towards the fish, beneath its belly, then up to just below its gills.

Bring your hand up with a fast, smooth action, scooping the fish clear of the water so that it lands on the bank.

you are unlikely to come across such still waters as a survivor. Your prey is more likely to be brown trout (and rainbow and other trout in North America) in clear running water. Like all trout, 'brownies' are very nervous fish. They spend a lot of time lying hidden, choosing places such as undercut banks, water-rat holes in stream and river banks, under rocks – in short, wherever there is a safe haven. Even though it is impossible to get anywhere near a trout in open water, you can catch them by hand, or 'tickle' them, while they are lying in cover.

Approach the area of cover very quietly, lie on the bank and put one hand in the water and wait until it is at or near water temperature. Then, starting at the downstream end of the cover, pass your hand gently and carefully through the water, working upstream. Touch the tail of the fish first. Stroke it gently a few times and then, still stroking the body, work slowly towards the gills. As you reach the gill area, grasp the fish quickly and firmly under the gills and toss it on to the bank. The general rule is to move slowly all the while you are tickling the fish, but to move like lightning once your fingers are beneath the gills.

To give yourself the best chance of success in tickling trout, observe the following points:

- Study the water carefully to identify bankside areas where trout will almost certainly be lying in cover. Watch where they head for when startled, or – this is more difficult – where they emerge from into open water.
- Tickling trout is usually a fairly tense business, but do your best to control any nervousness while

FISH TO AVOID

In general, fish provide very good food, but you should be aware that poisonous species exist. These occur in tropical waters, especially in lagoons and around reefs and atolls. Avoid fish from these waters with any of the following characteristics: an odd shape compared with species of the northern temperate region — for example, a box-like appearance; spines or thorns; an absence of scales; a slimy covering; sunken eyes; horns; glossy gills; an unpleasant smell; horny or protruding lips; irregular rows of teeth.

Do not eat any eggs, gills, intestines or livers of any fish from tropical waters. They contain poisons which are not destroyed or weakened by even the most thorough and prolonged cooking.

feeling for the fish with your fingers. If they are not relaxed it will be difficult to avoid an involuntary start when you first touch your prey. This will alarm the fish and scare it away.

POISONING FISH

In some of the warmer parts of the world native peoples make use of various plants which contain substances poisonous to fish but harmless to humans. An example is the derris, found in South-East Asia. The powdered roots of this plant, thrown into a stream above a pool, will render its occupants unconscious, and on surfacing they can be collected.

The crushed seeds of the barringtonia, a tree growing near the seashore in Malaya, have a similar effect. Lime will kill fish when introduced into streams. It can be obtained by burning fragmented limestone, chalk, coral, bones or shells. If more fish are obtained by any of these methods than can be eaten at one time, be sure to clean and dry them for future use.

13

NAVIGATION

In everyday life our navigation skills are rarely extended. When driving our own car we can rely on a comprehensive system of road signs and even on-board computers to direct us. When being transported we can depend on another person to know the way. In a survival situation, however, navigation usually proves much more difficult and may well be a matter of life or death. Therefore it is of prime importance to the survivor to know what means are available for regaining safety.

In fact in most survival situations navigation skills will be essential. For to find his way back to safety will be the main aim of the survivor, even when he is in a moderately hospitable environment with food and shelter to hand. Navigation becomes even more important if the survivor is in a remote location with little of either commodity, so that he will be unable to maintain health and strength for long.

MAP-READING

The ability to use a map and compass is the bedrock on which navigation skills are best built. Naturally, if the survivor has a grounding in this art and has the necessary tools with him, direction-finding should not present great difficulties. However, the survivor may be unable to map-read and for this reason the basic skills, as well as other navigational techniques, are described in this chapter.

It should be borne in mind, however, that in many survival situations, both military and civilian, neither a map nor a compass is available. In the first case, this is often because the survivor is an escapee and has had all his navigational aids confiscated by his captors. While the civilian is more likely to have retained his map and compass, in some cases it is precisely because he has, for some reason, become separated from these that he is lost. If you find yourself in either situation, you will need to fall back on other navigational tools.

MAPS

Since maps come in various scales and unfolded sizes, to meet the requirements of different situations, it is important to make your selection with care. Normally soldiers are issued with an Ordnance Survey map drawn to a scale of 1:50,000, whereas pilots receive maps with a much smaller scale and therefore covering a far greater area. In some war situations special survival maps are issued, and these are normally printed on cloth.

If you are a short-term survivor in a situation that involves armed conflict, the sound of fighting should guide you in the right direction, and in any case you may well still be in possession of your map. If, on the other hand, you have escaped captivity and have no map, work out the rough direction to follow and look for a map on the way. This suggestion is not as fanciful as it sounds, for maps are available from a great variety of sources, although unfortunately many of these – for example, vehicles, telephone boxes, schools and libraries – present danger in that you must go near or into populated areas. For the soldier, dead troops are another source of a maps, but here it is particularly important to exercise caution in case the enemy is still in the vicinity.

SETTING MAPS

Assuming you already had a map with you or, alternatively, have obtained one, the first thing to do is to 'set' it – that is, orientate it to reflect what you are looking at and, by extension, to determine your position. To do this, look out for a distinctive and permanent landmark – a river or a church tower, for example. (Note, however, that the shape of a wood, field or still water is difficult to discern from the ground.) If you can locate this landmark on your map, it should be easy

> ### Survival tip
> Although I have drawn a line on the map shown on page 232, it is for demonstration purposes only, and it is not good policy to mark your map in any way. In the SAS, to do so is an unforgivable crime. Not only does annotation of any kind make the map hard to use at a later date, or by someone else, but, more seriously, if it comes into enemy hands it can reveal a great deal about your movements.

For normal map reading use this type of compass. Make sure that the needle settles quickly and there is no bubble in its housing.

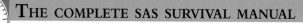

Setting a map in relation to the land is simple. Turn the map to align identifiable features with those on the ground. It is best to use linear features such as roads, rivers and railways, or prominent places such as towns, villages or mountains.

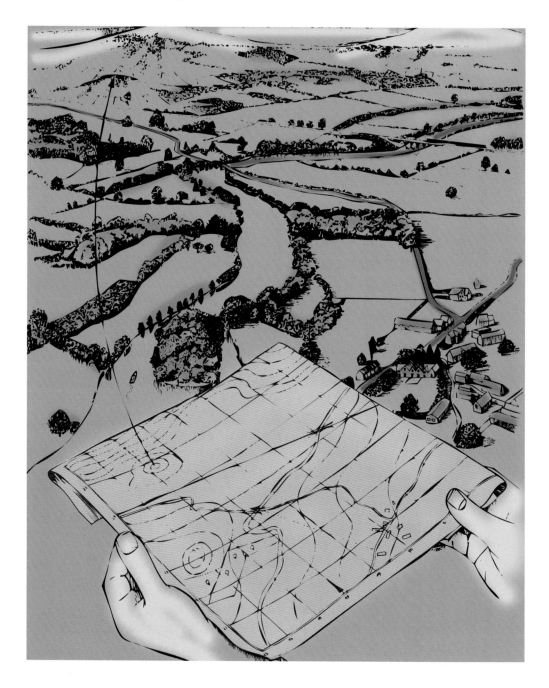

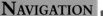

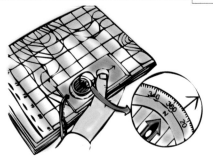

enough to align the map so that it shows the feature in the orientation in which you see it.

Alternatively, if you have a compass you can use this to set the map. Lay the compass flat on the map along a north—south grid line. Turn the map and compass together until the magnetic compass needle points north.

To set a map with a compass, turn it until the needle lines up with a line of latitude.

USING A COMPASS

To find your position with the help of a compass and map, first point the compass at a prominent object on the ground that you can see and identify on the map. Hold the compass steady and turn the compass housing until the magnetic needle lines up with north. The bearing to the landmark can then be read off the housing dial. For example, if the first bearing is 5700 mil, less the magnetic variation (*see* page 234), which we calculate at 40 mil, adjust the compass dial to 5660. Now place the top right-hand compass edge against the prominent point and pivot the whole compass until the magnetic needle is pointing north. Draw a line.

Repeat this procedure for a second prominent object. If that bearing is, say, 0650 mil, adjust for magnetic variation 0610 and plot as before. Where the two lines meet is your position.

Point the compass at a prominent feature. Turn the housing until the needle aligns with the north indicator, and read off your bearing to the feature from the rim.

IMPROVISED COMPASS

An improvised compass was described in an earlier chapter (*see* page 36) and there follow further ideas on improvisation. It is possible to make a compass by magnetizing a needle, nail, pin, razor blade, or indeed any suitable metal object (it must react to a magnet) – provided it can be suspended so as to swing freely as a pointer. This pointer can be magnetized by stroking it with one pole of a magnet, but it will require re-magnetization from time to time. Stroke the pointer repeatedly in the same direction, using the same end of the magnet each time.

Sometimes, suspending the pointer can present

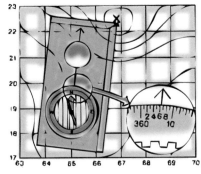

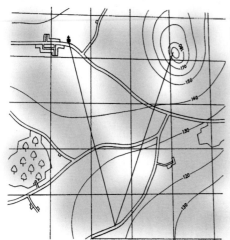

Transfer two or more bearings to your map. Where the lines intersect will be your position.

difficulties, especially if the thread is stiff or has been twisted. This will inhibit free movement of the pointer. If using a small object such as a needle or a scalpel blade, the problem can be overcome by pushing the pointer through a small piece of cork, two or three matchsticks, or a small twig. It can then be floated in still (not salt) water. Take great care that the water container itself is not made of any metal that affects the pointer magnetically. The pointer will now swing freely. It is also possible to magnetize a metal compass pointer using electricity, if you have a low-capacity battery (*not* a car battery) producing more than 6 volts and a length of insulated copper wire. For low voltages like these,

MAGNETIC VARIATION

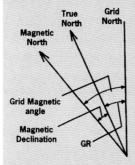

If you plan to use a compass, it is important to understand that, while most people are familiar with the concepts of True North and Magnetic North, there are in fact three 'Norths'. True North is the fixed geographical location at the North Pole, this north is not normally a factor for consideration in matters of navigation. The less well known North is Grid North, which is indicated by the vertical lines on a map. It is this North, in conjunction with Magnetic North, with which the compass user is most concerned. Magnetic North is a strong magnetic attraction to which our compass points. Unfortunately for the navigator, this magnetic force is not constant, and the

compass needle will be subject to its variations. To compensate for the discrepancy between Grid North and Magnetic North, it is necessary to calculate the annual change. The map will show the date when it was printed and the direction of magnetic variation occurring at that time. (It is important therefore to use an up-to-date map, if possible.) This variation, which is normally small, is added to or subtracted from the Grid North to provide a more accurate bearing. A simple guide to what sort of adjustment to make comes in the form of a mnemonic:
Mag to Grid: get rid
Grid to Mag: add.
In fact, during my military career I have never used the magnetic variation – for the simple reason that it is normally so small in relation to the other errors that creep into map-reading as to be negligible.

THE SIX-FIGURE GRID REFERENCE

The map is covered with a network of light-blue grid lines positioned to represent a distance of one kilometre. Vertical grid lines (running north—south) are referred to as eastings, and horizontal grid lines (running east—west) are referred to as northings. A grid square is defined by reading the number where the lines cross at the bottom left-hand corner of the square. The illustrated grid square reads: 1729.

To locate a specific point with reference to a map, the normal practice in both military and civilian map-reading is to give a six-figure grid reference. To do this, first take the number of the easting (vertical line) to the left of the feature (in the example given here, a church) to be grid-referenced — in this case, 17. Next mentally divide into tenths the square in which the point occurs, so that a point halfway up or across the square is represented by the number 5. Assign a number between 1 and 10 to the point in question, and add this to your easting. You now have half your grid reference: 175.

Now repeat the operation in a vertical direction by noting the number of the northing (horizontal line) below the point (in this example it is 29) and add to the number between 1 and 10 that most closely represents how far above this line your point occurs — here it is 7. You now have the horizontal and vertical components, which, when combined, form the six-figure grid reference of the point at which the church is to be found: 175297.

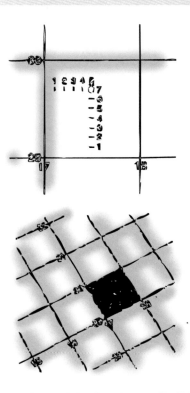

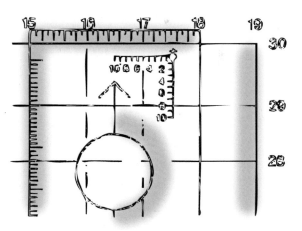

Sometimes compasses have built in romers, a scale along one edge that can be used to accurately divide a grid square up into tenths. Using this it is possible to give an exact grid reference without having to estimate any readings.

Survival tip
It is worth recalling that every radio set – right down to the smallest pocket receiver – has a permanent magnet as part of its loudspeaker. A radio can therefore be of use in improvising a compass. The successful survivor will always be on the lookout for ways in which items such as this can be used for original and sometimes unexpected purposes.

the insulation will often be a coat of varnish only. Wire of this type will be found in the coils inside radios, generators and other electrical equipment which form part of most types of transport.

Wrap the wire around the pointer, making as many turns as your wire allows. Connect up to the battery and allow the current to flow for 15-30 minutes. The North indicator will be the end of the pointer nearer to the negative battery terminal. (Remember, N = Negative and North.)

The basis of any compass is a magnetized pointer. Any steel needle, pin or razor blade can be magnetized by stroking it with a magnet or placing it inside a coil of wire connected to the terminals of a battery.

GLOBAL POSITIONING SYSTEM

Although its sophistication excludes it from being a part of survival navigation in the strict sense, a revolutionary piece of navigation equipment is worth mentioning here. Over the past few years navigation has taken a quantum leap, with the result that nowadays most Special Forces and pilots are issued with Global Positioning System (GPS) navigational aids. GPS was developed in the USA by the Department of Defense. A network of twenty-four military satellites orbit the earth. These satellites continuously relay information about the satellites' position. By using a receiver to

acquire information from several of the satellites it is possible to fix your position and orientation at any given point on the earth's surface.

The accuracy of the receivers depends on the efficiency of the receiving device. In fact, the system has a deliberate error built into it, referred to as selective availability or SA. This dithers the signal so that only a coarse acquisition (CA) is available. Accuracy under CA reception is about 40 metres (42 yards). In military use the so-called P Code is normally used with the receivers, overriding SA to give accuracy to 10 metres (11 yards) or even less. P Code receivers are very expensive. In times of heavy military activity, SA is switched off, allowing all receivers pinpoint accuracy.

The GPS works by searching the sky for satellite signals, which it locks on to immediately it finds them. A minimum of four such signals will prove satisfactory for the GPS's purposes, although a greater number will improve its accuracy. The information from the satellites is then computed into the form required by the user – to provide, for example, a Grid Reference, a Longitude and Latitude, or a height above sea level. Most modern receivers can be programmed to meet individual requirements, either at sea or on land.

The receiver can calculate an individual's position by a technique called satellite ranging, simple measuring his position in relation to a set of known objects. It will continually update his position, providing track and speed information, while he is moving. It will also record – without the need for a landmark – waypoints for future use.

My researches have turned up two good systems that are currently available in the UK: the Silver model and the Garmin 40. While both come with excellent instructions, it is essential with all such equipment to get out into the field and learn about its use in a hands-on way. Finally, don't forget your humble compass, since even the best GPS can shut down temporarily. For some, another consideration might be that GPS units have a huge appetite for batteries.

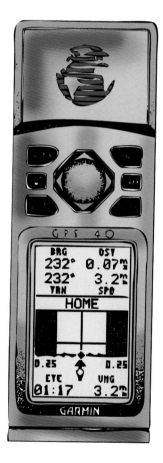

A modern GPS navigational aid is about the same size as a mobile phone. Although this is a very powerful piece of technology, it is always wise to carry a compass as a backup.

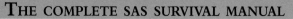

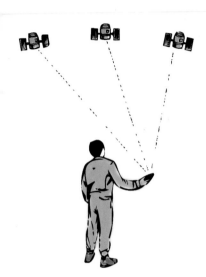

GPS navigation aids do not rely on the earth's magnetic field. Instead they measure their position relative to any four of a constellation of Navstar satellites that constantly circle the globe.

NAVIGATION WITHOUT A COMPASS

Although a compass provides the simplest and most foolproof way to establish your direction, there are many other methods you can employ if you are without one. In some cases these require no more than brain power and a few sticks and stones.

SOLAR NAVIGATION

One of the oldest ways to determine direction is the 'stick and stone' method, which is used as follows:

1 Cut a stick about 1 metre (40in) long and push it into the ground so that it stands vertically. Choose a very level spot where the stick will cast a shadow.

2 Using a small stone, mark the very end of the

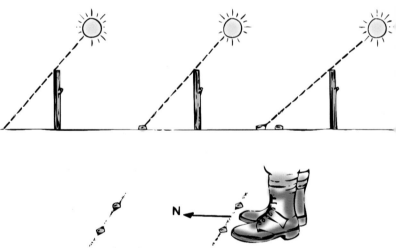

If the sun is shining, true north can be found using a simple stick and stone method. Marking the end of the stick's shadow with two stones, placed at the beginning and end of a twenty-minute period, gives a line that runs from east to west. Obviously, north is perpendicular to this line on the other side to the sun.

shadow as accurately as possible.

3 Wait for 15-20 minutes. You will observe that the shadow has moved. Mark the tip of this shadow with another stone.

4 Draw a line on the ground through the first stone and on through the second. This line will run from west to east.

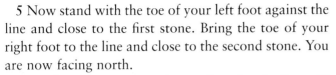

5 Now stand with the toe of your left foot against the line and close to the first stone. Bring the toe of your right foot to the line and close to the second stone. You are now facing north.

Note that the accuracy of your direction-finding will depend on the accuracy with which the shadow ends are marked, and the care taken in placing your toes to the line. A line drawn at right angles to your east–west line will produce a north–south indicator. Having established these cardinal points, you will find it easy to calculate any other desired direction by carefully dividing the space between them.

NAVIGATION WITH A WATCH

It is possible to determine direction by using a wrist-watch, and this need only be an average model, not a sophisticated chronometer. For the northern hemisphere the procedure is as follows. Check that your watch is set to local time and then point the hour hand towards the sun. Greater accuracy of aim can be achieved by placing a blade of grass or a thin twig so that it casts its shadow along the hour hand. A north—south line will now run from the centre of the watch face through the point on the dial edge which is midway between the hour hand and twelve o'clock. The northern indicator will be the end of the line further from the sun.

In the southern hemisphere the numeral 12 on the watch face should be pointed at the sun. The north—south line will run from the centre of the dial to a point on its edge which lies midway between 12 and the hour hand. The north indicator will be the end of the line nearer the sun. Again the watch must be set to local time.

Should you be in any doubt as to which end of the line is pointing north, remember that the sun is in the eastern part of the sky before noon, and in the western part in the afternoon. If you are facing north, the morning sun will be on your right-hand side and the afternoon sun will be on your left.

Provided that you have an analogue watch (one with hands) and the sun is strong enough to cast a shadow, it is possible to find north. In the Northern Hemisphere, point the hour hand at the sun. The north-south line bisects the angle between this hand and twelve o'clock.

Northern hemisphere

Southern hemisphere

In the Southern hemisphere, align twelve o'clock with the sun instead.

ASTRAL AND LUNAR NAVIGATION

The stars (and planets) have been used as aids to navigation for many centuries. They still have an important role to play in this area, and are also used in map-making. All study of the stars is fascinating, but for the survivor it may prove to be of vital importance.

The brightest stars appear to be set in fixed patterns, known as constellations. It is true that individual constellations do not vary, either in their particular patterns or in their relationship to one another. However,

Learning just a few of the more prominent constellations will help your night navigation. If you use the stars for direction, but intend to move by day, always leave an accurate marker in the ground to show the direction of travel.

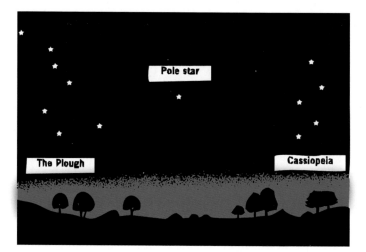

a brief observation will reveal that the whole body of stars seems, because of the earth's rotation, to be wheeling around a central point.

In the northern hemisphere the central point of this wheel is marked by Polaris, which is also called the Pole Star or North Star. Because of its central position it shows no apparent movement, remaining static in the sky above the North Pole. It follows, therefore, that if you can locate Polaris, you have an accurate indication of True North.

The best guide to Polaris is the constellation known as the Plough or the Big Dipper. This is made up of seven main stars. The two stars furthest from the Plough's 'handle' always point towards Polaris. If you

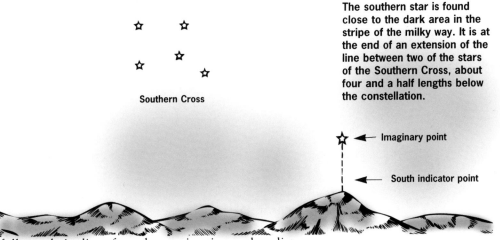

Southern Cross

The southern star is found close to the dark area in the stripe of the milky way. It is at the end of an extension of the line between two of the stars of the Southern Cross, about four and a half lengths below the constellation.

← Imaginary point

← South indicator point

follow their line for about six times the distance between them, you will find Polaris, even though it is not a particularly bright star.

You can confirm the accuracy of your star identification if you know that Polaris lies halfway between the Plough and another constellation called Cassiopeia. This consists of five main stars forming a 'W' shape, one half of which is slightly more flattened than the other. It appears on the other side of Polaris, almost opposite the Plough, the two constellations seemingly balancing each other. Neither the Plough nor Cassiopeia ever sets if seen from any country lying north of 40° N latitude.

In the southern hemisphere there is, unfortunately, no star that corresponds to Polaris – that is, no star which

The moon can also be used to indicate south, But this is also not a particularly accurate method. Imagine a line between the horns of the moon. The point where it intersects with horizon marks the north-south line.

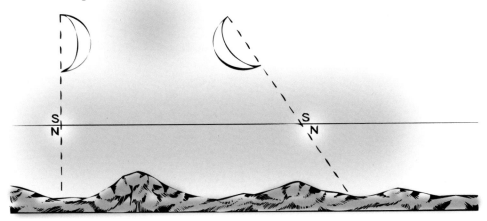

> **Survival tip**
> In a non-military situation it is advisable to mark your direction of travel so that any would-be rescuers can locate you. Leave a message of your intentions giving the time and the date.

remains fixed above the South Pole. Look instead for the Southern Cross. This is made up of four main stars, with the fifth, a fainter star, showing a little off the centre of the cross. Follow a line through the longer of the Cross's members for four and a half times its length. This brings you to the point where such a south pole star would be. Note a landmark directly below this point, as this point will indicate south.

To check that your line is extended in the correct direction, note that it should pass through a group of four faint stars soon after leaving the Southern Cross.

Note that if you are determining your direction ready for the next day's travel, you should take the bearing by using a stick laid on the ground as a pointer. This will continue to show the direction to be followed in the morning.

LUNAR NAVIGATION

There are two methods you can use to navigate by the moon which will produce reasonable results.

The first method is based on the quarter-moons. Produce a line through the horns of either of the quarter-moons down to the horizon. Where it meets the horizon it will indicate south, if you are in the northern hemisphere. It will indicate north if you are in the

DIRECTION INDICATED

Local time	First Quarter	Full Moon	Last Quarter
1800	South	East	
2100	South-west	South-east	
Midnight	West	South	East
0300	South-west	South-east	
0600	West	South	

southern hemisphere. Tests have demonstrated that this can provide a rough but very ready guide whenever you are moving at night.

The second method employs the quarter-moons and the full moon. If you have been able to set your watch to local time, or check that it is already correct, you can obtain a good indication of direction from these phases of the moon. Use your watch in conjunction with the table on the opposite page.

STAR MOVEMENT NAVIGATION

If the sky is partly obscured and you are unable to locate and identify individual constellations, the stars can still be employed for navigation. You can get a general indication of direction by making use of the knowledge that they appear to be wheeling around the sky. Detect which way any star is moving and you will know roughly which direction you are facing. Two fixed reference points are needed to detect star movement. A reliable way of establishin these is to set up two sticks (see diagram above) in the ground so that they are like the sights of a gun aimed at any prominent star you can see.

If the star appears to be:
• Looping flatly towards the right, you are facing south (approximately)
• Looping flatly towards the left, you are facing approximately north
• Rising, you are facing approximately east
• Descending, you are facing approximately west.

OTHER NAVIGATION METHODS

In addition to navigation by observation of the stars, the sun and the moon, there are other methods particularly suitable for the survivor lacking a map and/or compass.

Vegetation

If you find yourself in an area which supports much vegetation, there are probably many navigational clues close at hand. Although clues provided by vegetation will not provide very accurate directions, they will give a rough indication of direction. Even when the sun is not visible, many wild flowers will track its position, particularly yellow open-cup varieties. As a result, around midday in the northern hemisphere they will be pointing south. However, you must take account of any wind that may be moving the plants in a particular direction.

   Plants strive to benefit from the sun's light, the only exception to this rule being the family of mosses. Preferring damp conditions, mosses grow best on those aspects of trees, rocks and other features which receive the least sun. These aspects are those that face towards the Poles. Mosses therefore give an indication of direction that is the opposite of that provided by the rest of the plant kingdom.

Prevailing winds

In desert areas, where local aids to navigation are usually very sparse, prevailing winds can be of great assistance in direction-finding. The wind shapes sand dunes, and therefore if you know what the prevailing wind direction is, or can in some way deduce it, you can obtain a very rough indication of direction by observing the shape that the dunes take in response to the wind.

TRAVELLING THROUGH HOSTILE TERRITORY

Assessment of your survival situation may have identified the need to travel to a safer, more favourable location or to a place from where the chances of being rescued are greater. It may be advisable, or indeed urgent, to undertake a long journey, and if so you should apply the same careful deliberation to your travel plans as to your initial assessment. The first task is to check your direction, using one or other of the methods described above, and the next is to check the weather.

WEATHER

If you are snug in a snow hole there is little point in leaving it if outside a 'white-out' snowstorm is blowing. Many survival books go into great detail

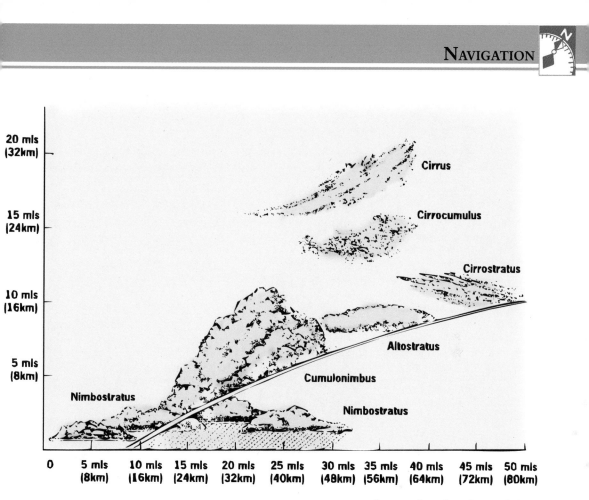

about the weather. I do not follow this line of reasoning, as weather forecasts get it wrong more often than they get it right. I have learnt, however, that watching the clouds gives some indication of the forthcoming weather. Since clouds are made up of water vapour, if there are a lot of clouds overhead rain is in prospect. This is particularly so in mountainous areas, where low, large clouds are forced to rise by the mountains and, in doing so, cool and produce rain.

CAMOUFLAGE AND CONCEALMENT

In a tactical situation remember your basic training in movement and camouflage. Camouflage your face and hands with charcoal, and check that your equipment is not rattling or making other tell-tale noises. Moving in daylight is best avoided in hostile country, but if you must do so, select a good route. Stay concealed and

As a survivor there is very little that you can do about bad weather, other than recognise its approach. One way of doing this is to observe the clouds. A clear sky with the odd very high cloud will indicate a good day. Low, dark clouds will indicate rain. Check the sky above your direction of travel; with a little practice you should be able to anticipate the weather conditions for several hours ahead and be able to prepare for it.

Make life a little easier for yourself when you are on the move by construcing a simple rucksack-style frame on which to carry equipment.

Even when the water is shallow, use a buoyancy aid or a stick to assist your crossing.

silent, and maintain a good overall direction. Make sure you spot the enemy before he spots you.

CARRYING EQUIPMENT

Check that you are equipped to provide food and shelter each time you camp, by taking with you what is really necessary, and what you can comfortably carry. Your ability to carry heavy loads over long distances will be greatly improved if you construct a sturdy carrying frame.

If you have a shelter sheet (top), roll your equipment into a bandolier pack (above). Carry it around your shoulder (left).

OBSTACLE CROSSING

One of the main obstacles you may meet is a large river. An evaluation of your fitness and the breadth, depth and strength of the river should help you to decide how to cross. Never cross a river unless you are completely confident that you can make the opposite bank. Check that the opposite bank is not so high that you cannot climb out. In shallow water where you can see the bottom, use a staff to assist your crossing.

In deeper water use a flotation aid, such as piece of wood. If you have made a good bandolier pack, this

will serve as a flotation aid. Another method is to remove your trousers and knots the legs.

If there are several of you in the party and a rope is available, cross individually by using a rope loop. This is secure as it allows two people to hold as the third crosses. It is recommended for fast-flowing rivers. However, for extremely wide or large rivers it may well prove more practical to construct a simple raft. You might consider at this stage whether travelling by river will assist you. You can use any material that will float, but be careful to make your raft strong enough to take some punishment. Secure your equipment to the raft with rope or string.

Water provides one the major obstacles when travelling. Even in shallow water use a buoyancy aid or stick to assist your crossing.
In deeper or strong flowing rivers, use a pendulum system to cross.
1 Loop a rope so that the first man crosses supported by the other two.
2 The second man crosses, supported by a man from either bank.
3 The third man is then brought across by the other two.
Use a long pole if you have a sick or injured person in the party. Make sure that he is in the middle and consider linking arms as well.

In a survival situation it is not wise to take any avoidable risks. Therefore only abseil if it is absolutely necessary.

Climbing, both up and down, is dangerous. Always make sure you have three points of contact at any one time.

MOUNTAIN AREAS

Unless you have some overriding reason to do so, you should not need to travel above the tree line. Mountainous areas are normally exposed and hostile, and if you can find an alternative route you should take it. Take care when walking on rocky ground, since a broken ankle will immobilize you. Should you find yourself in mountainous terrain consider the following:

- You can descend by using a rope. Make a site harness and lower yourself down gently.
- Climbing up is sometimes easier than climbing down. If you have to climb, send the best climber first, on a rope that he can secure for the less able. If you are alone with no rope, make sure that you have at least three points of contact with the rock at all times. Do not put your knees on the rock, since it will unbalance you.

14

SEARCH
AND RESCUE

SIGNALLING

SEARCH PROCEDURES

If you find yourself isolated in a hostile environment, it is inevitable that you will pin your hopes on being rescued before too long. Whether or not this happens depends on three crucial factors: where you are; how long you have been missing; and your ability to contact potential help.

The first factor is relatively straightforward: you are more likely to be found, or to be able to make contact with others, if your location is not remote. The second is not as clear-cut. Although you may succeed in surviving for some time while people are trying to find you, a protracted failure to locate you will eventually force the search party to conclude that there is no hope and give up. The third factor is one over which you have a large degree of control. When people are looking for you, perhaps risking their lives to do so, it is up to you to play your part by sending out signals as effectively as possible.

SIGNALLING

One of the most important priorities in any survival situation is communication. Clearly, if it is at all possible you must let the outside world know that you need help. Signalling for help takes two forms: Direct Communications and Ground Signals.

DIRECT COMMUNICATIONS

Rapid rescue from hostile environments requires a radio capable of sending location and distress signals. Ideally this should be capable of automatic operation and manual transmission, as this gives the Search and Rescue vehicles or aircraft something on

Use your radio sparingly until rescue is at hand – to save the battery.

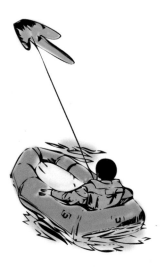

Survival tip

Civilians do not normally have access to high-tech military equipment, but in many cases you can find a civilian equivalent. A good example is the mobile phone as an alternative to radio communications. When I was testing new survival equipment with RAF SAR, the chief winch man told me that some of his most successful rescues had been carried out by people leading them directly to the site by using a mobile phone.

which to home in. Radios such as the Sarbe-5 and Tacbe BE499 are well suited to sending and receiving signals in this situation and are compatible with the system known as SARSAT (Search and Rescue Satellite-Aided Tracking).

LO-CAID

A fairly new piece of equipment, used by the military but also available on the civilian market, LO-CAID launches an emergency location balloon. The hand-held cylinder is contained in a plastic waterproof container. When activated, it inflates a special, radar-reflective helium balloon which rises into the air on a tether to a height of 33 metres (100 feet). Because of its aerodynamic shape, the balloon will stay aloft for five days, even in very strong winds. The balloon can be detected at distances of up to 39 kilometres (24 miles) by radar and 19 kilometres (12 miles) by sight.

GROUND SIGNALS

The term 'ground signals' embraces a variety of forms of signalling. In all cases the visibility and range of such signals depend on the location of their source and the weather conditions. Ground signals can be used in

Radar can detect a Lo-caid balloon at distances of up to 39km (24 miles).

Survival tip

Strobe lights are used in civilian settings – notably in discotheques – but for survival purposes a more readily available alternative is the camera flash. This blindingly bright and highly visible light source, either incorporated in the camera or a separate item of dedicated equipment, has been used successfully in a number of dramatic mountain rescues.

HAND SIGNALS

1 Yes
2 Pick me up
3 All well

1 No
2 Can proceed
3 Drop equipment

Do not try to land

By day, make smoke. At night, burn a bright signal fire.

addition to radio communications, to increase your chances of detection by potential rescuers, but in the absence of such equipment they are the sole means of signalling distress.

SIGNALLING FIRE

One of the most readily detected ground signals is that produced by an outdoor fire. A fire specifically intended for signalling to potential rescuers should be prepared and kept ready for lighting at the first signs of help arriving (*see* page 146). During the day it is best to make a fire that produces a large volume of smoke, while at night a bright flame without billowing smoke to obscure it, is required. If time and fuel allow, build three fires in a triangle.

HELIOGRAPH

Using reflections produced by the sun is an excellent way of signalling for help. This can be done by using any mirrored surface – for example, the wing or rear-view mirror of a vehicle, a vanity mirror or even the mirror from the toilets of a crashed aircraft.

However, by far the most effective device of this kind is a purpose-made heliograph. The word means 'writing with the sun' and the function of this mirror device is to deflect the sun's rays so that they reach a rescue aircraft or boat with a high degree of accuracy. A new, one-handed heliograph has recently come on to the market, costing only a few pounds, and with a little practice it is possible to achieve great accuracy with it.

Tests have proved the Mayday signalling mirror accurate at distances of over 32km (20 miles). Usefully, it also floats.

TORCH

At night even a small hand torch can be seen from a great distance, both on land and at sea, given the right weather conditions. It is important to conserve your batteries for such use.

STROBE

The strobe light has been used for military purposes for many years. Basically a flashing blue or white light, it is visible for many kilometres, especially at night. The strobe works well as a ground signal in both closed terrain such as jungle and in open mountainous areas. Conserve the batteries until rescue is imminent.

FLARES

Provided they are seen, all fired flares, irrespective of the colour, will be investigated by the relevant agencies. Once fired, a flare will reach a height of around 90 metres (300 feet) and is likely to be visible within a radius of many kilometres, depending on the weather

Fire flares only when a rescue party or an aircraft is in your vicinity. Use white flares at night or over dark forest areas, and green or red flares by day or over snow.

conditions. When snow is on the ground or falling, you should fire a red or green flare rather than a white one.

Small packs of survival flares can be purchased from good camping stores or chandlers. Care should be exercised with their use as they are a potential weapon. Always follow the instructions to the letter and aim the flare skywards.

Warning: Any flare pistol is dangerous and should be handled only by a competent person.

WHISTLE

Not only is the mouth whistle a very simple signalling device that rarely breaks down; it is also one that you can use as much as you like since it does not require batteries or any fuel other than breath. Almost all lifeboats and life-jackets carry whistles, and at sea they are very effective for locating people in the water, or those who are lost at night or in foul weather. Likewise, any properly planned survival kit will contain a whistle. A recent innovation from Canada is a new kind of whistle which produces a very high pitch and has a range which, it is claimed, exceeds one kilometre (1100 yards).

Various ground hand signals can also be used and will be recognized. A few of these basic Rescue signals are illustrated on page 252.

A Perry whistle comes attached to almost every life-jacket. To signal your presence, give six blasts per minute with one-minute intervals between series of blasts. The reply is three blasts.

SEARCH PROCEDURES

Throughout the world, Search and Rescue (SAR) teams work in similar ways. For this reason the information given below, which is based on current British practice, will be relevant to such operations wherever you happen to be lost. By understanding these widely adopted procedures you will have some idea of how long it is likely to take for trained rescuers to arrive.

If it is known only that you are missing, and your location has not been determined, the SAR teams will search using a set of standard procedures. The area covered will be based on the best estimated overall

guide of your last known location. How the search is carried out will be determined by the size of the area to be covered, the nature of the terrain, the weather and operational necessity and the availability of SAR teams.

If you have followed the correct procedures and notified others of your intended route, or you can establish radio communications, then what is known as a contact search will be initiated. Designed to concentrate rescue efforts on a smaller area, this kind of search accelerates the process of making contact and generally achieves a positive result faster than a broader-based operation.

MOUNTAIN RESCUE BY HELICOPTER

Helicopter crews can take a considerable amount of time assessing the possible problems that may be encountered in attempting a mountain rescue. In very windy conditions, for example, it is not uncommon for a pilot to make several attempts to hover close enough to the casualty to be able to get a winch man or mountain rescue team in position. Once he has arrived at a workable 'hover', his next priority is to assess the safest method of rescuing the casualty.

To ensure that no important aspect of the situation is

The search pattern is largely determined by the subject's last known location and the terrain. In jungle areas the main watercourse is followed. In mountains there is a search of each peak. A search using aircraft and foot parties is common on land. At sea a ship often holds the search course while aircraft sweep out to the flanks.

Normally the helicopter rescue pilot will select a safe place to land. If you have place marker panels, make sure they are secure. Do not approach the helicopter until signalled to do so by the pilot or the air crew. When it is safe to do so, approach from the pilot's front right side. Rotor blades are deadly, so stay well clear. Tilt your upper body forward and keep your head down, but stay observant.

overlooked, RAF helicopter rescue crews use a standardised system of assessment and briefing. The following priorities normally govern decision-making:

1 Aircraft safety
2 Winch man safety
3 Survivor safety.

It may seem odd that the casualty is at the bottom of the list of priorities. The thinking behind this practice is that there is little justification for endangering the lives of several rescuers unless the situation absolutely compels it. Helicopter rescue – like other forms of rescue – is concerned first of all with risk assessment.

The person or persons waiting at the incident site to be rescued by helicopter should advertise their presence as effectively as possible. If a Sarbe, Tacbe or other radio is available, it will be possible to direct the SAR helicopter to the site verbally. An alternative is Dayglo orange or pink plastic panels measuring about 60cm (24in) square. Seen well from the air, and even more visible when moved in a generous rhythmic motion, these can prove a useful item in the kit of anyone risking a survival situation.

At night flash a torch, but if a helicopter approaches, direct the beam at the ground. The reason for doing this is that the crew will be using night-vision goggles to improve safety and even small external light sources shone directly at them can seriously reduce their efficiency or stop them working altogether. It is possible, however, to signal with lights without affecting the efficiency of night-vision goggles. Firefly rescue strobe lights in particular have proved to be very effective for this purpose.

Secure any loose kit before the helicopter arrives in the hover. Even with care, some downwash from the rotor cannot be avoided and winds of 80kph (50mph) can be generated. Also, if the pick-up site is very steep or otherwise dangerous, try to prepare a belay – an anchor point – for the winch man's use.

Strop winching is carried out at sea or when the rescue is dangerous for the winch man and a double lift is not possible. As the strop reaches you, place it over your head and arms, then tighten the loop. Give the thumbs up to indicate your readiness to be lifted and then lean back in the harness as you rise. Warning: it is advisable to let the strop touch the ground or water to discharge any static electricity generated by the helicopter.

15

SURVIVAL
AT SEA

USING A DINGHY

WATER

Should you be unfortunate enough to find yourself in an emergency situation on the high sea, your chances of survival will largely be determined by whether you are actually in the water or in a dinghy or life-raft.

In the first case, survival time will normally be determined by the temperature of the sea and the wind, as well as by the clothing you are wearing and the safety devices available. The first two factors are a matter of chance. However, people do not just find themselves in the sea. If you end up in a dinghy, unless you were very unlucky you will have been provided with a lifebelt or a lifebuoy before or after entering the water.

You may need to use smoke and distress flares if you are shipwrecked. They are dangerous, so when they are burnt out drop them into the sea – well out of harm's way.

USING A DINGHY

If you can swim and there is a ship, land or safe harbour in sight which you feel you can reach by doing so:

- Do not discard any of your clothing unless it is impeding you and thus reducing your safety.
- Always swim with a steady breast-stroke, and avoid overexertion.
- If you have a life-preserver you will find it less tiring to swim or float on your back.
- In a real emergency, taking your trousers off and tying the bottoms of the legs will provide a brief flotation aid. Holding the waistband, throw them over your head, trapping air. This is a very useful survival measure, and is worth practising before an emergency arises.

If, as the result of an incident at sea, you are adrift in an inflatable dinghy, your chances of survival are much greater than if you are at the mercy of the water. Not only are you protected to some extent from the sea's chilling and exhausting effects, but most modern dinghies are equipped with survival equipment and water. However, since it may be some time before the rescue services are able to reach you, it is wise to take the following precautions:

- Check the dinghy to make sure it has inflated correctly, since it may be faulty as a result of having been stored for many years.
- Close the cover of the dinghy to maximize your protection from the weather.
- Do not jump about or move clumsily in the dinghy

If your dinghy has capsized, use the straps or rope on the vessel to pull it the right way up. This not an easy task in a heavy sea. Once you are inside the dinghy, erect the storm cover and check for leaks. If there are any, they can be plugged with the stick-on patches provided.

The single-seater aircraft life-raft is cramped and uncomfortable when used for any length of time, especially in rough seas. On the plus side it will be fully supplied with rescue equipment and rations.

LOCATION

If you find yourself in the sea, it will be because your ship or your aircraft has gone down. Even without an SOS call most such incidents are noted by others, either by direct sighting or by radar. Search and Rescue operations will be implemented quickly and you can expect help within a short time. To assist in the search all means of signalling at your disposal should be activated as soon as possible.

since it can easily be damaged.
- Check the survival supplies on board and the contents of your pockets.
- Prepare any available signalling equipment for use.
- If there is more than one person in the dinghy, check the physical and mental condition of each occupant and plan to attend to any particular problems when the priority of preparing the signalling equipment has been addressed.
- Do not remove any of your clothing at this stage.
- If you have been forced to abandon your boat or aircraft as a result of an accident, make sure your dinghy does not drift into any sharp debris.
- Bale out any seawater.
- Set up a solar still if one is carried on board (*see* opposite page).
- Make a survival plan, taking into account the condition of any occupants of the dinghy who have been injured or are unduly distressed.
- Stay happy – you could be in the water.

WATER AND FOOD

When you are forced to survive at sea for any length of time, your first priority must be to obtain drinking water. If there is time before taking to the dinghy, drink as much as possible. A person in an average state of health can live for about 25 days without food, but without water 10 days is the best you can hope for. Surprisingly, a human being can survive on as little as 112g (4oz) of water a day, but even so any water on board must be rationed. Drink no water the first day, since your body will have retained a certain amount, then, depending on the supply, start off with about 280g (10oz) each per day. Keep this up if possible for the first four or five days, then slowly reduce the daily intake to around 112g (4oz). Always moisten your lips and mouth before swallowing.

The most important thing to remember about water if you are in a survival situation at sea is: never drink seawater. If you do, it will sharpen your thirst instead of relieving it. The salt accumulation in the body will then lead to increased body-fluid loss and eventually to kidney failure.

Survival time with no water at all varies according to the weather conditions and your state of health, but approximately four days of deprivation will result in derangement, and death will follow after seven to twelve days.

For this reason it is extremely important to follow a survival routine in which you do everything possible to conserve your existing body fluids. To this end, cover yourself in order to protect yourself from the sun and wind. Also, avoid unnecessary effort and try to keep calm. All these measures will minimize sweating. In addition, wet your clothing occasionally with seawater as this will help you to keep cool.

If there is a solar still on board the dinghy, put it into action as soon as you can. The device uses the heat of the sun to produce drinking water. It takes some time to produce even a small amount but is nevertheless a useful item of survival equipment.

RAINWATER

When you have addressed the major priorities of preparing signalling equipment and administering first aid, make plans to collect and store rainwater. Use every item that can be pressed into service. Anything that can be used to collect water should be cleaned in preparation for use, and the method of collection should be decided. If necessary, clean pieces of cloth can be used to mop up water so that it can be wrung out into a container. In an extreme situation the cloth can be sucked dry instead.

Carry out a test run of your water collection set-up and make any necessary adjustments, since it is imperative not to lose any water you could have collected. If all storage space is full, drink all you can, for your body is also a storage vessel.

FOOD

In the adverse conditions of survival at sea, water is a major priority. However, food will become a matter of urgency after a few days as lifeboat and life-raft rations are designed to meet only your energy requirements and will not be adequate as sustenance. Survival rations usually come in two forms: dry biscuits or boiled sweets. Both are carbohydrate-based, and their consumption will demand little water intake.

FISHING AT SEA

The most readily available source of supplementary food is fish, and therefore all lifeboats hold a fishing kit. Using a survival fishing kit is not very different from fishing in fresh water. However, there are certain guidelines that should be followed in a marine survival situation.

1 The hook is an indispensable part of all fishing tackle. However, when you are adrift in a rubber dinghy it is even more important than when you are on dry land to bear in mind the damage its point can do.

2 Handle fish with care since even non-poisonous species have extremely sharp fins or gill covers.

3 Hooks will be in short supply, so handle them with care and make sure they are tied securely to your line.

4 If you use a knife to clean fish, make sure that the blade is folded away when not in use. Secure your knives to your clothing or to the boat.

5 Cut surplus fish into strips and sun-dry them for future use.

6 A dinghy will often attract a large number of small fish. Use a bailing bucket to scoop them aboard. As well as sustenance, they can be used alive or dead as bait for bigger species.

7 Most species of sea fish are edible, palatable and nutritious. However, be aware that some, especially in the

Rainwater is a valuable resource. Collect as much as possible, by using the widest area you can construct. Even try to get your clothes wet – they can be wrung out into your mouth.

SOLAR STILL

There may be chemical or osmoration kits in the dinghy which will purify seawater. If so, follow the instructions for use. A solar still for distilling seawater may also be provided in the dinghy. If so, you should put it into operation immediately.

DESALINATION

You must never drink seawater — it will cause death more quickly than drinking no water at all. It contains so much salt that the kidneys are unable to cope with it and will ultimately fail. And do not drink urine, as it will cause illness. However, both liquids can be treated in the solar still to produce drinking water (see page 261).

An alternative way to make seawater potable is to use a desalination pump. The BCB Survivor 06 LMCM, the world's smallest such device, is an ideal means of producing water on a lifeboat or raft. With a pumping rate of forty strokes per minute,

it will produce 1.1 litres (1.9 pints) of fresh water from seawater per hour.

The pump obtains drinkable water by reverse osmosis, and uses no fuel. When seawater is placed under pressure against a fine membrane inside the pump, the larger salt molecules are unable to pass through; neither are any impurities. The molecules of desalinated water pass into the collection chamber. The membrane will give approximately 4000 hours of service. Do not use the pump if petroleum-based hydrocarbons are on the sea's surface, to avoid damage to the membrane.

tropics, are poisonous. A main characteristic of these fish is the lack of scales of the kind seen on other species. Many are encased instead in bony spines or a box-like shell. Avoid all jellyfish and sea snakes.

MARINE HAZARDS

Exposure to the elements and dehydration are the major threats to your survival when lost at sea. Unfortunately, there are additional threats to contend with. In the northern hemisphere, the dangers posed by the sea are lessened if you are in a dinghy. But should you find yourself in tropical waters then there is the risk of sharks, barracudas and swordfish. However, while there is a slight possibility that predatory fish will attack the dinghy, there is little chance that they will harm you, despite what the horror movies suggest, since most are cowards and will retreat if given a thump on the head.

Most large fish will, in fact, be attracted to the dinghy by its size, initially seeing it as a source of food. But provided you do not panic, and make sure that

If you have a soluble shark repellent on board, spread it on the water and manoeuvre your vessel to the centre of the inky patch.

everyone is safely in the vessel with no limbs exposed, your visitors will soon lose interest. Do not try to tempt them to move away by throwing tit-bits into the water, as this will have the opposite effect and they will become more curious. Above all, remember that most man-eating fish live in deep waters and rarely hunt on the surface.

You may have a shark repellent in the dinghy's survival kit which works by clouding the water with an inky chemical. It also produces an acrid smell that the shark dislikes. This substance has been used successfully by the British Navy in the Indian Ocean, where sharks abound. The trick is to position your vessel in the centre of the cloud for maximum safety.

SEASICKNESS AND OTHER CONDITIONS

Apart from exposure to extreme weather conditions and dehydration, the most serious of the marine hazards is seasickness. Not only does it make the survivor feel terrible but it also causes continuous vomiting which reduces body fluids rapidly. Most dinghy survival packs contain a small pack of seasickness tablets. However, if these are not available and there is sufficient room, sufferers should lie flat in the dinghy and keep as warm as possible.

Salt water can also affect the eyes and exposed skin if the survival situation is prolonged. Excessive exposure to bright sunlight and the reflections produced by the sea is also a major hazard. Parched lips and cracked skin can may cause some discomfort but will do little long-term damage.

16

SELF DEFENCE

Situations arise in which soldier and civilian alike must defend themselves. Evil people exist in most societies, and occasionally the most aggressive of these confront us. But unprovoked violence does not occur solely in pressured urban environments. A soldier, separated from and striving to rejoin his comrades-in-arms, will most likely need to defend himself. A civilian survivor should also be prepared, since in the developed and undeveloped world alike hostility is often the reaction to those seen as intruders or even simply as strangers.

BASIC RULES

Unfortunately, aggression and strength are the two factors which usually decide the outcome of any conflict, but in some cases surprise and confidence act as a deterrent to an assailant. Protecting yourself against physical harm is a natural reaction, but it is not enough just to rely on instinct.

To ensure your safety you need confidence in your ability to defend yourself. The relevant skills, and the confidence to use them appropriately, can be learned. The aim of this chapter is to teach them. Its contents derive from what is known in the military world as CQB – Close Quarter Battle. CQB is not a martial art – which is a quite different approach to self-defence – but is simply hand-to-hand conflict, whether it be one to one, or one against many.

In most societies it is reasonable to use such force as is deemed necessary to get yourself out of trouble. Any less may expose to you to danger, while any more may land you in prison. Always be careful about what you use to defend yourself. For example, if you happen to have a hammer in your hand, do not use it on your attacker's head. Such actions are excessive if a blow to the kneecap will suffice. Carry weapons only for legitimate, lawful reasons – for a survival enthusiast in the

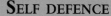

wild, a large Bowie knife is not out of place; in a city street its carrier would find himself in a lot of trouble.

FITNESS AND DIET

Another important rule of self-defence concerns your own health. Your body is fragile, therefore you need to take great care of it if it is to serve you well. This is important for everyone, but for a person who might find himself in a survival situation, particularly if facing the possibility of attack, it is essential. Good, consistent exercise, coupled with a healthy diet, keeps your body machinery operating efficiently. One of the major keys to effective self defence is speed. The swift action needed in such a situation only comes if the body is fit and the thinking is quick – fortunately exercise also sharpens the mind.

Any form of exercise you undertake should be tailored to suit your physical condition and any personal disability. In order to achieve the right balance, it is recommended that you seek advice from your doctor before embarking on any programme of exercise to which you are unaccustomed. Likewise, do not adopt a diet of your own devising unless it has been approved by a qualified dietician or a book with the backing of reputable practitioners in this field.

FEAR

When you are faced with a threatening situation, fear is nature's first line of defence. It sends adrenalin surging through your body, bringing your senses alive and alerting them to danger. But fear can also render you incapable of action, and the only way to counter this reaction, which increases your vulnerability, is to takes steps to control it. Learning one or two basic CQB moves will help stop you freezing with fear when a crisis arises. As you go about your daily life, get into the habit of thinking about what would you do if something threatening suddenly happened. If you find yourself getting frightened, try breathing in and counting to

> **Survival tip**
> Even before I joined the SAS, I knew that to join that highly selective regiment's ranks I would need to be completely fit. I reached the required state of fitness simply by running and walking. As a result, the amount of air that was forced into my system, and the ease with which it was absorbed, made me quick and clear-minded. To this day, walking and the occasional downhill jog are still my favourite methods of staying fit.

ten, then breath out counting to ten. Repeat this pattern three times and you will feel a lot better. The combination of anticipating trouble and controlling your breathing will allow you to master fear, so that you can act wisely and promptly when faced with real danger.

AVOIDANCE TACTICS

The situations I describe below are ones that occur with growing frequency and could happen to anyone. Because they are easily imagined, they are a good teaching aid. Having learned how to deal with such situations, you can apply this expertise to the context of surviving attack in the wild.

To take the first example, if you are travelling on an underground train and a bunch of drunken football hooligans enter your carriage, you can simply get off at the first available opportunity, thus avoiding any confrontation. Since self-preservation is your aim, as it is for all of us, the simple fact is that you have won. You have not allowed yourself to be drawn into trouble and the certainty of injury.

As a second example, if you are confronted by a bunch of thugs intent on kicking your head in and you manage to outrun them because you are fitter, again you have won. I have had my fair share of fights, but I freely admit that there have been times when I knew instinctively that it was better to turn and run.

The lesson that these examples teach us is simple: avoid trouble or get away from it. Of course there will be times when you have no option but to fight. In this connection I recall the very first words that the CQB instructor said to my class when I joined the SAS. Never fight unless you have to, he said. But, if you have to, win – no matter how. The survivor fully committed to surviving (not just hoping blindly to come through unscathed) will understand these two complementary approaches to self defence – fight or flight – and know which to use in any given situation.

Before we examine the techniques of physical self

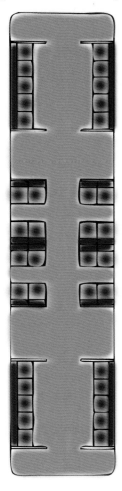

On public transport, many tense situations can be avoided by thinking ahead. Choose your seat carefully, and move if you do not feel comfortable. Sit near an exit and get off at the next stop if you see trouble brewing.

defence, let us consider the non-violent alternatives, for there is much that you can do before you are forced to resort to violence.

- Always give yourself an escape route. If you cannot find one – create one.
- If the person(s) who confronts you is only after money it may be best to give it to him.
- Talk – you may diffuse the situation and will certainly be creating more time for help to arrive.
- Stay as calm as you can – in particular, ignore verbal abuse and resist the temptation to get angry.
- Try to anticipate any attack by the person confronting you. If an attack seems imminent, shout as loudly as possible. This the time to activate a personal alarm if you are carrying one.

Warning: The techniques described here, which are based on Close Quarter Battle procedures, are intended only for self defence. Care should be taken not to escalate the situation by using physical force beyond that which is necessary. Furthermore, restraint should be shown when practising any CQB technique with a partner as it is easy to inflict unintentional injury.

TARGET AREAS OF THE BODY

If evasive action is out of the question and you are forced to defend yourself by attacking, it is important to know what can be regarded as the target areas of your (would-be) assailant's body.

Hair If your assailant has hair long enough for you to get hold of, it can provide a good means of controlling him. With a firm grip on someone's hair, you can, in most cases, force him into a submissive position.

Eyes Being very vulnerable, the eyes are an obvious target, but causing blindness or permanent damage definitely needs to be justified. However, there are ways of causing temporary lack of vision which will allow you time to escape. Two of these are explained below.

Nose The nose is a very effective place to hit. A hard

If you are forced to fight, think of your assailant's weak areas. Consider his hair length, whether or not he is wearing glasses, and note those parts of his body protected by thick or padded clothing.

upward blow will cause the eyes to water and produce a temporary loss of vision. Making the nose bleed, both externally and internally, will interfere with the breathing. You can afford to give an assailant's nose a hard blow without fear of too much damage.

Ear Slamming the ear with the palm of the hand will produce shock waves down the delicate ear channel, while hitting both ears at the same time can result in unconsciousness. If you are in close conflict with the assailant, try blowing hard down his ear.

Temple A sharp blow to the temple, directly between the eye and the ear, will cause loss of vision and dizziness. As with the ears, a double-sided blow will greatly increase the effect.

Throat Like the nose, the throat is a good place to hit hard. There is little chance of doing serious damage, but it will have the effect of interrupting the air supply and make breathing difficult.

Stomach Although the stomach is a very soft and sensitive area, in the case of most adults in good health it can sustain a lot of punishment. If the assailant intends to fight, his stomach muscles will go tight and the chances of winding him are slight. But a good blow, when he is off guard, will serve to make him double up.

Crutch This is a very tender area, especially in male assailants. Even a moderate blow to the testicles will have the desired result. A kick to this area is possible both from in front and from behind. If you find yourself in a tight struggle, get your hand firmly on your assailant's testicles and twist. Better still, sink your teeth into them. I guarantee that either will work.

Thigh Despite the fact that strong muscle surrounds the upper leg, it is still a tender spot. Use your knee or your toe, but if you are using any form of weapon other than your own body, this is the place to strike.

Knee Any damage to the knee will cause a great deal of pain. The blow should be downward, or directly against the kneecap. As with the thigh, if you are using a weapon this is a good place to strike since it will be an effective deterrent, but there is no need to worry about inflicting a serious injury.

Foot There is far less muscle protection in the foot (and the hand) than in most of the rest of the body. For this reason striking it forcefully or stamping on it will produce painful results.

THE BODY'S WEAPONS

In a situation where no other weapons are available, you will have to defend yourself with the body's own arsenal of natural weapons.

Fist To strike someone we naturally ball our hand into a fist and in most cases this is the best way to use it. Use your fist to jab at pinpoint areas such as the chin, nose and stomach. A useful refinement on this is to fill your hand with loose change – the added weight produces a much heavier blow.

Open palm The open palm can be used two ways: back- handed, against the temples, or to slap the ear. If your assailant is of light build, consider using a vigorous slap.

Heel of the hand Striking with the heel of the hand can be very useful where a quick, sharp, upward jab is required, such as a blow to the chin or to the back of the neck. You are less likely to injure your hand with this technique than if you punch with the fist.

Elbow You need to be either side on or have your back to your opponent to use your elbow effectively. However, from these positions it is one of the most vicious of the body's weapons. Jab the elbow into your assailant's stomach or the small of his back, or, if you are on your knees, into his testicles.

Knee Although it is a very strong weapon, the knee is unfortunately limited in its action to the lower part your assailant's body. It is also, owing to its heavy muscle coverage, the slowest of the body's weapons. Nevertheless, this battering ram can cause severe and lasting damage, and in a serious situation it may have to be used.

Toe A hard kick can be used just as readily, and to as

good effect, as a blow from the hands. The only difference is that your balance must be taken into consideration. Unless you are practised, keep all kicks below waist height, where they will in any case do more to immobilize your assailant.

Heel The heel is an excellent natural weapon when you are attacked from behind, or when you are trying to free yourself from an attacker. While the reach of the heel is limited, it can be used on the lower body without fear of doing permanent or serious damage.

Teeth Sinking your teeth into any part of your assailant's body will produce severe pain and deflect him from his intended actions.

THE BODY'S DANGER AREAS

The intensity of your response to being attacked will be determined by the gravity of the situation. In other words, the greater the danger the harder you will fight back. Nevertheless, it makes sense to exercise caution when hitting an assailant's head. Unless the threat is particularly extreme, it is hard to justify causing death, blindness, deafness or any permanent physical or mental damage by hitting this vulnerable area.

USEFUL WEAPONS

There are a number of items that can be used for self-defence without causing serious injury to the assailant.

Newspaper A rolled-up newspaper or magazine can cause discomfort and put an assailant off balance when jabbed into his face.

Pen A ball-point pen stabbed into his nose will make any attacker think twice about renewing the assault.

Dirt or sand A handful of dirt or sand thrown into your assailant's face will temporarily blind him. Act fast, so that he does not have time to cover his eyes.

Umbrella An umbrella or a walking stick extends your reach, so that you can maintain some distance between yourself and your attacker. This tactic is particularly useful against an assailant armed with a knife.

Your own body is equipped with many natural 'weapons', but they can be augmented by some everyday items. For example, a rolled-up newspaper or magazine can be jabbed in the face, causing considerable pain.

A fistful of sand thrown in an assailant's face gives you time to make an escape.

LETHAL WEAPONS

Do not under any circumstances carry a potentially lethal weapon such as a gun or a knife. Very few people would actually use either, and if your assailant receives the impression that you are not prepared to use the weapon in your hand, it is not a deterrent. Conversely, if you do use a gun or a knife yet fail to stop the attack, because violence breeds violence the same weapon may well be used against you.

COMBAT TECHNIQUES

If you have decided to fight your assailant, the only weapons at your disposal will be those we have considered above and the speed of your reactions. If time permits, discard any outer clothing that may impede you. (The same applies if you have decided instead to run, since the aim in both cases is to achieve maximum speed of movement.)

ON-GUARD POSITION

It is important to begin your self-defence in the right way – in terms of both the physical resistance you offer to your assailant and the psychological effect it will have on him, for he will be more cautious if he sees you have been trained in effective defensive strategies and are prepared to use them. Adopt the standard on-guard position by standing facing your opponent, with your feet apart and your left foot slightly forward. Bring your right arm up, with the palm of your hand flat and a few inches from your mouth. Extend your left hand out at about chest height.

Try to flow when moving; maintain your balance. Keeping your balance will allow you to defend and strike more precisely

In a frontal attack, maintain your stance and let your attacker come to you.
1 Block with the left arm, and punch up to his chin using an open palm.
2 Push the chin completely back to unbalance your assailant, then bring your knee up into his groin. Once contact is broken, run.

when the opportunity arises. Most attackers rush in, hoping to pin their prey. Prevent this happening with quick footwork or defensive arm jabs. If you intend to kick, keep your shots low, since they will cause more pain this way and they make it easier for you to main-tain your equilibrium.

FRONTAL ATTACK

Keep your balance, lean forward and use your left hand in a sweeping motion to open up your assailant's body to attack. Drop your head slightly and at the same time jab the heel of your right palm under his chin. Forcing his head and shoulders back with your hand will leave his stance exposed. To finish off this tactic, bring your knee up into your attacker's groin. Break and run.

ATTACK FROM THE REAR

Bend your head forward and then bring it back sharply to administer a reverse head-butt. At the same time, ball your fist and strike backwards at the groin. If the attacker has an armlock on you, he may be persuaded to release it if you use your teeth or bend one of his fin-gers back very hard.

INDEX